Climate and Society

Climate and Society

Transforming the Future

ROBIN LEICHENKO
KAREN O'BRIEN

polity

First published in 2019 by Polity Press

Reprinted 2020 (twice)

Polity Press
65 Bridge Street
Cambridge CB2 1UR, UK

Polity Press
101 Station Landing
Suite 300
Medford, MA 02155, USA

ISBN-13: 978-0-7456-8438-3
ISBN-13: 978-0-7456-8439-0(pb)

A catalogue record for this book is available from the British Library.

Library of Congress Cataloging-in-Publication Data
Names: Leichenko, Robin M., author. | O'Brien, Karen L., author.
Title: Climate and society : transforming the future / Robin Leichenko, Karen O'Brien.
Description: Cambridge ; Medford, MA : Polity, 2019. | Includes bibliographical references and index.
Identifiers: LCCN 2018046807 (print) | LCCN 2018051795 (ebook) | ISBN 9780745684420 (Epub) | ISBN 9780745684383 (hardback) | ISBN 9780745684390 (pbk.)
Subjects: LCSH: Climatic changes--Social aspects.
Classification: LCC QC903 (ebook) | LCC QC903 .L45 2019 (print) | DDC 304.2/5–dc23
LC record available at https://lccn.loc.gov/2018046807

Typeset in 9.5pt on 13pt Swift
by Fakenham Prepress Solutions, Fakenham, Norfolk NR21 8NL
Printed and bound in the United States by LSC Communications

For further information on Polity, visit our website: politybooks.com

Contents

List of Illustrations

Figures

Tables

Boxes

Acknowledgments

We received tremendous support in writing this book from students, collaborators, family, friends, and colleagues. In particular, we thank the undergraduate students in Robin Leichenko's Climate Change and Society course at Rutgers University and Karen O'Brien's Environment and Society course at the University of Oslo for their patience, suggestions, and good humor as we tested different topics and ideas in our lectures and seminars and discovered what resonated with them, challenged them, and frustrated them. We also thank them for their recommendations about how to make the chapters more accessible, relevant, and meaningful from a student's perspective. We hope that the resulting book is engaging and empowering, and that it inspires both critical reflection and transformative action.

We are grateful to Emma Longstaff (formerly at Polity) for her early encouragement and support of the project and to Jonathan Skerrett at Polity for guiding the book to completion. We thank Coleen Vogel and numerous anonymous reviewers for valuable comments on the original prospectus and draft manuscripts. At Rutgers, Khai Hoan Nguyen gave valuable feedback on every chapter and pilot tested many of the book's illustrative examples when she served as a teaching assistant in the course. Former PhD students Peter Vancura, Adelle Thomas, Mark Barnes, and Ally Sobey helped with development of the course syllabus and lecture materials. Robin also benefited from many discussions with current and former graduate students, colleagues, and collaborators, including Katya Bezborodko, Mike Brady, David Eisenhauer, Ana Mahecha Groot, Melanie McDermott, Katy Ryan, and Julie Silva. At the University of Oslo, Ann Kristin Schorre, Linda Sygna, and Leonie Goodwin provided research assistance and support, and seminar leaders Irmelin Gram-Hanssen, Gail Hochachka, Milda Nordbø Rosenberg, and Sadik Qaka contributed ideas, feedback, and suggestions on the course material and text. Karen is also grateful to researchers Kirsten Ulsrud and Morgan Scoville-Simonds for chapter comments, and to Dan Jesper Lagerman and Sasha Stoliarenko for valuable feedback on the manuscript. Karen also benefited from discussions and ideas shared by PhD students in the Oslo Summer School in Comparative Social Science Studies courses on Climate Change Adaptation and Transformations towards Sustainability.

Finally, we thank our colleagues in the Department of Geography at Rutgers University, the Rutgers Climate Institute, and the Department of Sociology and Human Geography at the University of Oslo for creating positive and supportive environments where new ideas and approaches can thrive. Most of all, we thank our families, Charles and Henry Strehlo and William Solecki, and Annika, Espen, Jens Erik, and Kristian Stokke, for their patience, love, and inspiration throughout every phase of this process.

Glossary

adaptation is an action taken in response to the impacts of climate change in order to reduce risks, losses, and damages or to take advantage of new opportunities.

adaptive capacity is a measure of the ability of a household, community, sector, etc. to respond to and recover from climate change shocks and stresses.

albedo is a measure of how much solar energy hitting the Earth's surface is reflected back to space.

Anthropocene is a proposed geologic epoch beginning in roughly the middle-twentieth century, which is distinguished by the pervasiveness of human influences on Earth-system processes.

anthropogenic forcing is a measure to describe the influence of human activities in altering the Earth's energy balance.

barriers to adaptation include cultural, technical, economic, institutional, and other factors that hinder efforts to plan and implement adaptation actions.

Capitalocene is the idea that capitalism is the central driver of human influence on global environmental systems.

carbon budget is an accounting of how much carbon dioxide can be added to the atmosphere to avoid exceeding particular global temperature targets.

carbon footprints are a measure of the amount of carbon dioxide released into the atmosphere as a result of the activities of individuals, groups, organizations, cities, or other entities.

carbon lock-in is the inertia associated with large-scale investments in fossil fuel energy and transport systems, which fosters continued reliance on fossil fuels and resistance to shifts to other energy sources.

carbon markets are systems for purchasing, trading, and selling permits to emit greenhouse gases; they are intended to create incentives to reduce emissions.

carbon sequestration is a process whereby carbon dioxide is captured from the atmosphere and stored over long periods.

carbon tax is a tax placed on fossil fuel producers and/or consumers in order to create an incentive to reduce consumption.

climate change fingerprint is evidence that climate change has increased the likelihood or magnitude of an extreme weather event.

climate fiction (cli-fi) is a genre of literature that includes novels and stories about climate change.

climate impact assessment is a type of research study that entails investigation of the social, economic, biophysical, and/or ecological consequences of climate change for a particular sector, industry, community, or region.

climate justice is a concept that draws attention to the inequalities across different communities, groups, and nations in contributions to rising greenhouse gas emissions and in vulnerabilities and capacities to adapt to the impacts of climate change.

climate normals are averages of temperature and precipitation in a region over a 30-year period.

climate resilience is the ability to recover from climate shocks and stresses and to take measures to reduce exposure, vulnerability, and risks associated with future climate events.

climate shocks and stresses are climate-related events such as heat waves, droughts, floods, or wildfires, which cause or may cause harm or damage.

climate skeptic is an individual who doubts the validity of the scientific evidence of climate change.

co-benefits are indirect, positive benefits that result from taking particular actions.

community-based adaptation is a collaborative approach to adaptation that engages community members with the goal of aligning adaptation decisions to the needs and cultural preferences of the community.

compound events are simultaneous or sequential extreme weather events that affect one or more locations.

congestion surcharges (also known as congestion pricing) entail taxes or tolls on motorized vehicles that are levied during periods when traffic is heaviest and roads are most congested with the goal of discouraging motorists from driving at those times.

consumption-based emissions are the greenhouse gas emissions associated with goods and services consumed within a particular country, including imported products that are manufactured or produced elsewhere.

co-production is the development of scientific information through two-way collaboration between scientists and members of a community.

cultural values are values, assumptions, and norms that are accepted and shared among a particular social group.

divestment campaigns are strategies that encourage institutions to reduce or eliminate their investments in the stocks of companies that produce oil, natural gas, and coal.

double exposure is a situation where a household, group, sector, or region experiences overlapping impacts or vulnerabilities as a result of exposure to both climate change and other large-scale economic and social changes such as globalization and urbanization.

drawdown is the idea that efforts to slow or reverse climate change should focus on maintaining carbon in the biosphere by both reducing greenhouse gas emissions and by pulling carbon out of the atmosphere.

dualistic view is a way of understanding humans and society as separate and distinct from the environment and nature.

ecological footprint is a measure of the total environmental impact of consumption and production in a city or region, including pressure on both surrounding local environments and global environments in the form of air and water pollution, land-use changes, and solid waste disposal.

ecosystem services is a concept used to quantify the benefits of the different aspects of the natural environment for human well-being.

El Niño–Southern Oscillation (ENSO) is a recurring pattern of warm and cool water temperatures in the equatorial Pacific Ocean which influences weather conditions in many other regions of the world.

emissions scenarios are projections of future levels of greenhouse gas emissions based on different sets of assumptions about population growth, energy usage, land use, economic growth, and other factors.

energy poverty is a situation where low-income individuals or households cannot afford to meet their basic energy needs for heating, cooking, and other uses without compromising their ability to afford food, medicine, shelter, or other basic needs.

environmental justice represent a concept that directs attention to the uneven spatial distributions of environmental hazards and amenities and, particularly, to the disproportionate exposure of poor and marginalized communities to environmental toxins and pollution.

externalities are costs, damages, or harms associated with particular actions, such as burning of fossil fuels, that are borne by society at large.

flexible adaptation pathways represent an approach to adaptation planning that provides decision makers with a wide range of options for adjusting strategies over time in light of future changes in climate risks.

food miles are an estimate of the greenhouse gas emissions associated with the total distance that a food item travels from the location where it is grown, harvested or raised to the location where it is consumed.

fossil fuel subsidies are direct and indirect payments and other forms of financial and regulatory incentives that governments use to support the extraction, production, transport, and sales of fossil fuels.

global climate models (also known as general circulation models) are complex mathematical models of the climate system which are used to make projections about future climates under different greenhouse gas emissions scenarios.

global production chains are associated with globalization and entail division of the production process for particular goods into separate components carried out in different countries.

global warming potential is an index used to compare how different greenhouse gases contribute to global warming based on how much energy each gas absorbs over a set time period (typically 100 years).

Great Acceleration is a period that began roughly in the middle of the twentieth century and continues to the present day and is marked by dramatic increases in consumption of fossil fuels, deforestation and land-use changes, greenhouse gas emissions, and other measures of human impact on the global environment.

green transitions are large-scale shifts away from fossil fuel energy sources towards alternative, renewable energy sources.

habits of capitalism are associated with high levels of consumption of energy-intensive goods and services which support and are supported by the globalized, capitalist economy.

human exceptionalism is the idea that humans are unique and categorically different from all other species on earth.

implicatory denial is the failure to integrate knowledge about climate change into everyday life and transform it into social action.

indigenous worldviews are a broad category of beliefs and cosmologies held by members of communities and nations that have territorial connections to particular regions and strong links to the environment.

inelastic demand is an economic term to describe situations in which consumer demand for a particular product or service is not responsive to increases or decreases in the price of that product or service.

information deficit model is a model of science communication which presumes that audiences need to better understand the science associated with an issue such as climate change in order to be convinced that the issue is a real and pressing concern.

instrumental value is a view that something has worth as a means to accomplish a desired end.

Intergovernmental Panel on Climate Change is an international body established in 1988 to regularly assess the state of scientific knowledge related to climate change.

intersectionality is the idea that forms of discrimination and marginalization tend to overlap across multiple categories such as gender, race, class, and ethnicity.

intrinsic value is the inherent worth of something in and of itself.

just transitions are a recognition of the equity implications of shifts to renewable energy with the goal of ensuring that these shifts do not increase poverty and inequality or cause displacement of polluting industries to poorer regions.

land grabbing is the increased external ownership and control of agricultural land within lower-income countries in Latin America, Asia, and Africa.

limits to adaptation are insurmountable social and biophysical obstacles that render adaptation actions ineffective.

low elevation coastal zones are low-lying land areas in coastal regions and island nations that are likely to be exposed to sea-level rise.

maladaptation is an adaptation action that contributes to increased emissions of greenhouse gases or to increased vulnerability of others.

market-based measures include taxes, subsidies, and other incentives intended to motivate changes in behavior of producers and consumers.

mitigation is an action taken to reduce greenhouse gas emissions associated with human activities in order to limit the rate and magnitude of climate change.

non-dualistic views are a way of understanding the relationship between humans and nature; they consider humans and human societies as inseparable or as one and the same with nature.

nudging is a term used in behavioral economics to describe approaches that help people to make the right decisions (e.g., those that contribute to emissions reductions) by making choices easy and obvious.

ocean acidification is the increasing acidity (decreasing pH) of ocean waters worldwide as a result of the absorption of carbon dioxide from the atmosphere.

paradigms are patterns of thought and ways of understanding systems and human–environment connections.

Paris Agreement is a 2015 agreement signed by more than 190 nations to limit climate change to below 2°C – and ideally to 1.5°C – by the end of the twenty-first century.

path dependency is a situation whereby construction of large-scale infrastructure systems such as road networks creates inertia for continued investments in those systems and discourages shifts to other types of systems.

peak oil is the maximum possible rate of extraction of oil on a global basis, after which available oil supplies and the rate of extraction are expected to permanently decline.

place attachment is an emotional and psychological connection to a particular location that contributes to a sense of individual and community well-being.

planetary boundaries are thresholds for the impact of human activities on the Earth's physical life-support systems, which, if crossed, will contribute to a heightened risk of abrupt and irreversible environmental change.

positivist science is an approach to scientific knowledge based on the application of the scientific method including evidence-based, empirical verification of theories.

post-normal science is an approach to scientific knowledge that recognizes that science is always developed within a social context and that science is neither objective nor value-free.

precautionary principle is the idea that precautionary actions should be taken on an issue where there is a plausible risk of harm, even if some aspects of the issue are uncertain.

prosumers are individuals and households that produce renewable energy (e.g., from rooftop solar panels) which is transmitted to the energy grid they also consume energy from the grid.

radiative balance is the relationship between incoming solar radiation and outgoing terrestrial energy.

radiative forcing is a measure of the influence that a particular gas has in altering the Earth's balance of incoming and outgoing energy.

REDD+ is a collaborative United Nations program that relies on financial incentives and other market-based measures to motivate farmers and landowners in less developed countries to reduce crop production or change their farming practices in order to reduce rates of deforestation and promote mitigation.

resource curse describes situations where exploitation of resources such as oil leads to neither economic benefits nor positive development outcomes.

root causes are economic and political processes, relations, and historical legacies that contribute to poverty, inequality, marginalization, and vulnerability of particular social groups.

self-transcendence is a mode of thought or consciousness that emphasizes serving a greater purpose, one beyond the needs of the individual.

sense of place is the meaning and emotions that are evoked in connection with a person's experiences and memories of a particular place.

social cost of carbon is an estimate of the combined cost to society from fossil fuel-related carbon emissions including, for example, costs associated with damage from extreme weather events, harm to public health, and reductions in agricultural productivity.

social norms are the unwritten ideas or rules of behavior that are considered acceptable by individuals within a particular group.

soft measures are policies and regulations that are intended to promote behavioral changes.

solastalgia is a feeling of sadness and distress that is created by recognition that climate change is impacting or harming an individual's home environment or another location to which an individual feels strong emotional attachment.

spatial displacement of emissions is the phenomenon whereby a country reduces its territorial emissions by increasing its consumption of imported goods and services that produce emissions elsewhere.

stranded assets are fossil fuel reserves that are owned or controlled by fossil fuel companies but are not extracted, creating the potential for large economic losses for these companies.

Sustainable Development Goals are 17 global goals adopted by the United Nations as part of the 2015 resolution "Transforming our World: The 2030 Agenda for Sustainable Development."

techno-managerial is a description of strategies for addressing climate change that emphasizes technological innovations, regulations, and policy changes.

territorial emissions are the greenhouse gas emissions associated with fossil fuel usage and other activities within a particularly country, city, or geographic region.

threat multiplier is the idea that climate change may contribute indirectly to the risk of conflict or social unrest because of its negative impacts on agricultural production, rural livelihoods, and ecological resources.

tipping points are thresholds where systems can shift irreversibly into another state or mode of operation with unpredictable and potentially dangerous consequences.

traditional ecological knowledge includes indigenous and related forms of knowledge about the relationships between living beings and the environment.

United Nations Framework Convention on Climate Change (UNFCCC) is an international environmental treaty which aims to stabilize emissions of greenhouse gases at a level that would "prevent dangerous anthropogenic interference with the climate system."

vulnerability is a predisposition or susceptibility to being harmed by a climatic event or circumstance.

vulnerability assessment is a type of research study that investigates why particular households, communities, groups, sectors, or regions are more susceptible to harm from the impacts of climate change.

vulnerability hot spots are spatial regions that are identified through vulnerability mapping as highly vulnerable to climate change.

1 THE SOCIAL CHALLENGE OF CLIMATE CHANGE

Setting the stage

Climate change is transforming the world as we know it. In some places, extreme or unusual weather events are raising concerns that "climate change is happening now." In other places, longer-term shifts such as increasing temperatures, melting glaciers, and rising sea levels are leading to existential questions about how climate change may affect future livability and survival. Along with increasing awareness of the reality of climate change, there is growing recognition that the sooner we take action, the lower the risks of severe, widespread, and irreversible global impacts (IPCC 2014a). But how should we respond? What can we do? Answering these questions begins with seeing climate change as more than an environmental issue. We need to look more broadly at social, economic, political, and cultural processes that are both driving climate change and influencing responses. We also need to look more deeply at how we see ourselves in the world, how we relate to others, and how our individual and collective decisions and actions are shaping the future for generations to come.

This book explores social causes, consequences, and responses to climate change. Our point of departure is that there are many different perspectives on the problem of climate change and many different approaches to solutions. When climate change is viewed as an environmental problem, the solutions are usually technical or behavioral, such as managing greenhouse gas emissions and promoting environmentally friendly lifestyles. When climate change is viewed as a social problem, the solutions expand to include economic, political, cultural, and institutional changes, some with the potential for transforming society in ways that address multiple global challenges, including poverty and inequality, food insecurity, biodiversity loss, and health crises. The United Nations **Sustainable Development Goals** (SDGs) for 2030 recognize that all of these global issues are linked, including SDG Goal 13: "Take urgent action to combat climate change and its impacts" (United Nations 2015).

This introductory chapter sets the stage for our investigation of the social and human dimensions of climate change. We begin by exploring the question of why climate change matters, what it means for economic, social, and natural systems, and its implications for issues of equity, ethics, and justice. We then consider climate change within the context of the **Anthropocene**, a new geologic epoch distinguished by human influence on Earth-system processes. We situate climate change in geological, historical, and future time frames and connect it to industrialization and ongoing globalization processes. We then explore openings and opportunities for reducing climate change risks and vulnerabilities. We conclude by making the case that climate change is a fundamentally transformative process; it is not only transforming Earth systems but also how we think about ourselves and our capacity to create change. We emphasize that future impacts and risks are not predetermined. There are enormous differences between a world that is 1.5°C warmer and a world that is 4°C warmer (New et al. 2011), and there are many openings and opportunities to create an equitable and sustainable future.

Why does climate change matter?

When asked why climate change matters, many people struggle to articulate its significance. Some feel that climate change is too abstract to really grasp, or that it "doesn't really affect me," or "isn't something that I need to worry about in my lifetime." Others see it as just one of many pressing social concerns, including global poverty, homelessness, disease, addictions, unemployment, terrorism, and military conflicts. With so many competing issues, it can be easy to downplay the significance of climate change or assume that technological innovations will eventually solve the problem. While many people are indeed very worried about climate change, they may be simply overwhelmed by its implications, convinced that it is too big an issue to address. All of these are examples of what Kari Norgaard (2006) refers to as **implicatory denial** – the failure to integrate knowledge about climate change into everyday life and transform it into social action.

Let's think about some of the many reasons why climate change does matter. Climate change matters for very practical reasons, as weather and climate are foundational to our everyday lives (the difference between weather and climate is discussed in chapter 2). When the weather becomes less predictable, we have to live with greater uncertainties and new risks. From mundane questions about whether to wear a jacket or carry an umbrella to decisions about the timing and location of outdoor excursions, festivals, sports events, and weddings, we manage our daily lives based on certain expectations about the weather and the climate. For some, the everyday experience of climate change means new inconveniences, such as navigating roads that are subject to more frequent flooding. For others, climate change involves adjusting routines, such as reducing daily water usage in response to long-term drought. For many, climate change means increasing exposure to extreme weather events, such as floods, forest fires, or heat waves. For all, climate change raises questions of what we value most and why.

Indeed, at a personal level, climate change matters because it affects things that we value. Most people care about access to fresh food, clean water, and good health. Some also care about experiences like skiing, ice fishing, or bird-watching. Many people value being able to work or exercise outdoors during summer months, or perhaps they value the experience of snow in winter. The diversity of plants, animals, and ecosystems matters, not just because of the benefits and services that they provide but because of the **intrinsic value** of nature. Values and identities will be influenced by climate change as it transforms conditions and experiences that matter to individuals, communities, and nations. Safety, prosperity, and a **sense of place** are among the many things people care about that are affected by climate change.

Climate change also matters because social and economic systems are organized around a preference for stability and predictability. Climate change alters fundamental rhythms of nature, and this has consequences for agriculture, fisheries, manufacturing, tourism, and many other sectors of the economy where activities depend on predictable seasonal patterns. For farmers, decisions on when to plant crops and when to apply fertilizer and irrigation are rooted in expectations about weather and climate. For fishermen, a warming ocean can affect what types of fish are available, when to catch them, and how much to harvest. For manufacturers, more frequent coastal storms can damage port facilities and disrupt global supply chains. For sectors such as ski tourism, a lack of winter snow can have dramatic repercussions for local economies.

Climate change means that the past can no longer serve as a reliable guide for the future. With baseline environmental conditions continually changing, water resource managers are facing difficult decisions about how to plan for future increases in demand. In sectors such as logistics, engineering, and construction, decisions ranging from where to site new warehouse facilities and how to plan new roads, bridges, and other major infrastructure projects are complicated by uncertainties about future temperatures, rainfall, flood heights, wind speeds, and sea levels. For the property insurance industry, extreme weather and a changing climate mean increased costs and less predictability about future damage payouts, which, in turn, means higher premium rates for their customers.

Climate change also matters for reasons of security and well-being. In addition to intangible losses to cultural identities and values, human populations are experiencing material damage to property and livelihoods, and in some cases they are being displaced from their homes because of sea-level rise, droughts, or floods. Climate change is a potential threat to national security in many countries, whether due to the disappearance of coastlines, stresses on agricultural and water resources, or population displacement and migration. In the future, it is likely to affect where people can live, what they can eat, and how they experience life. For this reason, young people, those who have (or anticipate having) children or grandchildren, and those concerned with the well-being of future generations have a critical stake in today's climate change policies and actions.

It is not only humans that will be affected by climate change. Climate change matters because it affects other species and ecosystems. For example, a warming and acidifying ocean influences marine ecosystems, affecting coral reefs as well as the entire food chain. Many species of fish feed on phytoplankton and zooplankton, which are sensitive to temperatures and ocean acidity. Some of these fish in turn provide food for other fish, for birds, and for mammals such as penguins, polar bears, dolphins, and walruses. Changing climate conditions represent an existential threat for some animals and plants, increasing the risk of local and global extinction. On land, the tolerance zones for plants have been gradually changing, with many species migrating north or south, or to higher elevations. Some tree species in temperate regions are threatened by pests and diseases that did not typically survive over winters when conditions were colder.

This brief overview shows that there are many reasons why climate change matters, and indeed climate scientists have identified numerous reasons for concern (see Box 1.1). However, as we will discuss in this book, climate change also introduces some reasons for optimism. It shows that the environment is directly influenced by human activities and that our actions can and do have impacts on larger systems. This implies that we have the potential to initiate positive changes that will contribute to a more equitable and sustainable future. Climate change matters because it shows that *we matter*.

Equity, ethics, and justice

In addition to the arguments described above, climate change matters for reasons of equity, ethics, and social justice. Both the causes and consequences of climate change are unequally distributed across society. Responses to climate change, including **mitigation** and **adaptation** policies, are also likely to have unequal consequences for different regions, sectors, communities, and social groups. Beyond documenting and cataloguing climate change impacts, vulnerabilities, and insecurities, there is a need to step back and think about the larger moral and ethical implications of climate change. One way to approach these issues is through the lens of equity. When we look at climate change through such a lens, some critical questions arise: Do individuals and countries who are doing the most to cause climate change have a moral obligation to help those who are most affected? What type of compensation should the victims of climate change receive, and from whom? Who is looking out for the rights of non-human species and future generations?

While the answers to these questions vary depending on philosophical perspectives, as well as beliefs, values, and worldviews, there is no doubt that climate change impacts and policies raise profound equity concerns (Klinsky et al. 2017). Equity, in its most general form, is a measure of whether something is fair to all concerned. A foundational equity issue raised by climate change is that those who are least responsible for causing the problem are most vulnerable to its impacts (O'Brien and Leichenko 2006). Not only are the poorest and most marginalized disproportionately affected, but climate change impacts can also exacerbate existing inequalities (Leichenko and Silva 2014). In addition to experiencing greater exposure to climate change, the poorest and most marginalized tend to have the least capacity to adapt and recover. These groups also tend to have little voice or political influence in processes and decisions that could potentially contribute to their security.

Box 1.1 "Reasons for concern" about climate change

The **Intergovernmental Panel on Climate Change** (IPPC 2014b) has identified the following five key reasons for concern to inform discussion about the risks of climate change.

1. Risks to unique and threatened systems
2. Risks associated with extreme weather events
3. Risks associated with the distribution of impacts
4. Risks associated with global aggregate impacts
5. Risks associated with large-scale singular events

These risks were evaluated by the IPCC based on projections of future global temperature change. In general, they show that the higher the temperature increases, the greater the risks. O'Neill et al. (2017) identify three additional metrics that can inform debates about long-term mitigation targets: rate of change; **ocean acidification**; and sea-level rise. These authors point out that these global reasons for concern do not explicitly account for differences in the exposure and **vulnerability** of socio-ecological systems over time and how they are influenced by societal conditions.

Is the IPCC's list of reasons for concern about climate change missing anything important? If so, what risks do you think should be added? Do you think this list makes an effective case for taking action to address climate change?

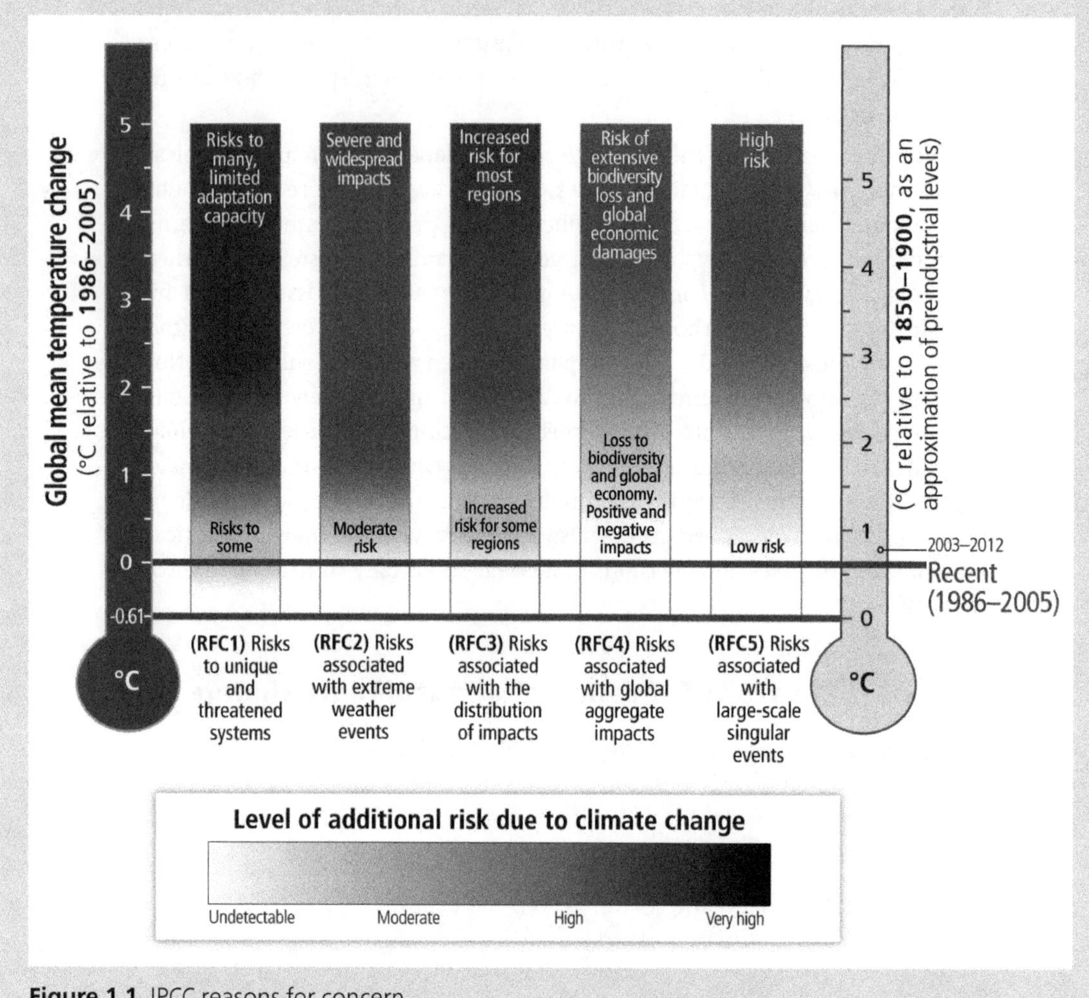

Figure 1.1 IPCC reasons for concern.
Source: IPCC 2014b

Beyond present-day impacts and responses, equity issues regarding climate change also relate to future generations. Gardiner (2006) describes climate change and its implications for future generations as a perfect moral storm, whereby its complexity provides a convenient excuse for current and successive generations to take weak and largely ineffective actions. Although it is sometimes easy to think of future generations as a distant abstraction, Davies, Tabucanon, and Box (2016) remind us that children born today will be living through a century of climate change impacts. As such, children and young people represent a critical link between current adults and future generations; they are potentially the strongest advocates for action on climate change (Flora and Roser-Renouf 2014).

The climate of the Anthropocene

Climate change is frequently described as "the greatest challenge for humanity." Yet it is important to recognize that human activities have long influenced the environment. From prehistoric times onward, activities such as hunting and fishing, gathering of fruits and seeds, and domestication of crops and livestock have affected the planet's biophysical systems. Human activities, such as making fires for cooking and heating or clearance of forests for agricultural production, led to changes in the composition of the atmosphere and the Earth's **albedo** (a measure of how much solar energy hitting the Earth's surface is reflected back to space), thus influencing the climate. Since the onset of industrialization, roughly 250 years ago, the scale and extent of human influence on the Earth's systems has become ever more evident and pronounced. The cumulative effect of human activities on the planet is now so pervasive that the present era is sometimes referred to as the Anthropocene (Steffen, Crutzen, and McNeill 2007). The Anthropocene is defined as an epoch where human activity has been the dominant influence on Earth-system processes (see Box 1.2). Although the onset of the Anthropocene is debated, it marks the end of the Holocene epoch, a time period lasting more than 11,000 years, during which environmental conditions supported the development and flourishing of human societies (Lewis and Maslin 2015; Zalasiewicz et al. 2017). The rapid and dramatic environmental transformations that characterize the Anthropocene have vital implications for human society.

With respect to climate change, the pervasive influence of human activity can be traced to the use of fossil fuel energy and the development of industrial societies. Prior to the industrial revolution, energy needs were met primarily by burning wood and other replenishable organic materials. Industrialization dramatically increased demand for energy to power industrial processes and transportation. This led to a rapid increase in the extraction and production of fossil fuels, first coal and later petroleum. Along with increased rates of logging and clearance of forests for agriculture, livestock grazing, and expansion of towns and cities, growing use of fossil fuel energy marked a turning point for human influence on planetary systems, including the climate system.

Demand for fossil fuel energy to run machinery, mine natural resources, and transport raw materials, finished products, and people accelerated with the expansion of industrial societies and the rapid growth of towns and cities after World War II. The period starting in about 1950 marks the beginning of what has been referred to by Steffen et al. (2015a) as the **"Great Acceleration"** (see Figures 1.2 and 1.3). Throughout this period, the increased burning of fossil fuels, along with deforestation and expansion of agricultural and livestock production, has contributed

Earth system trends

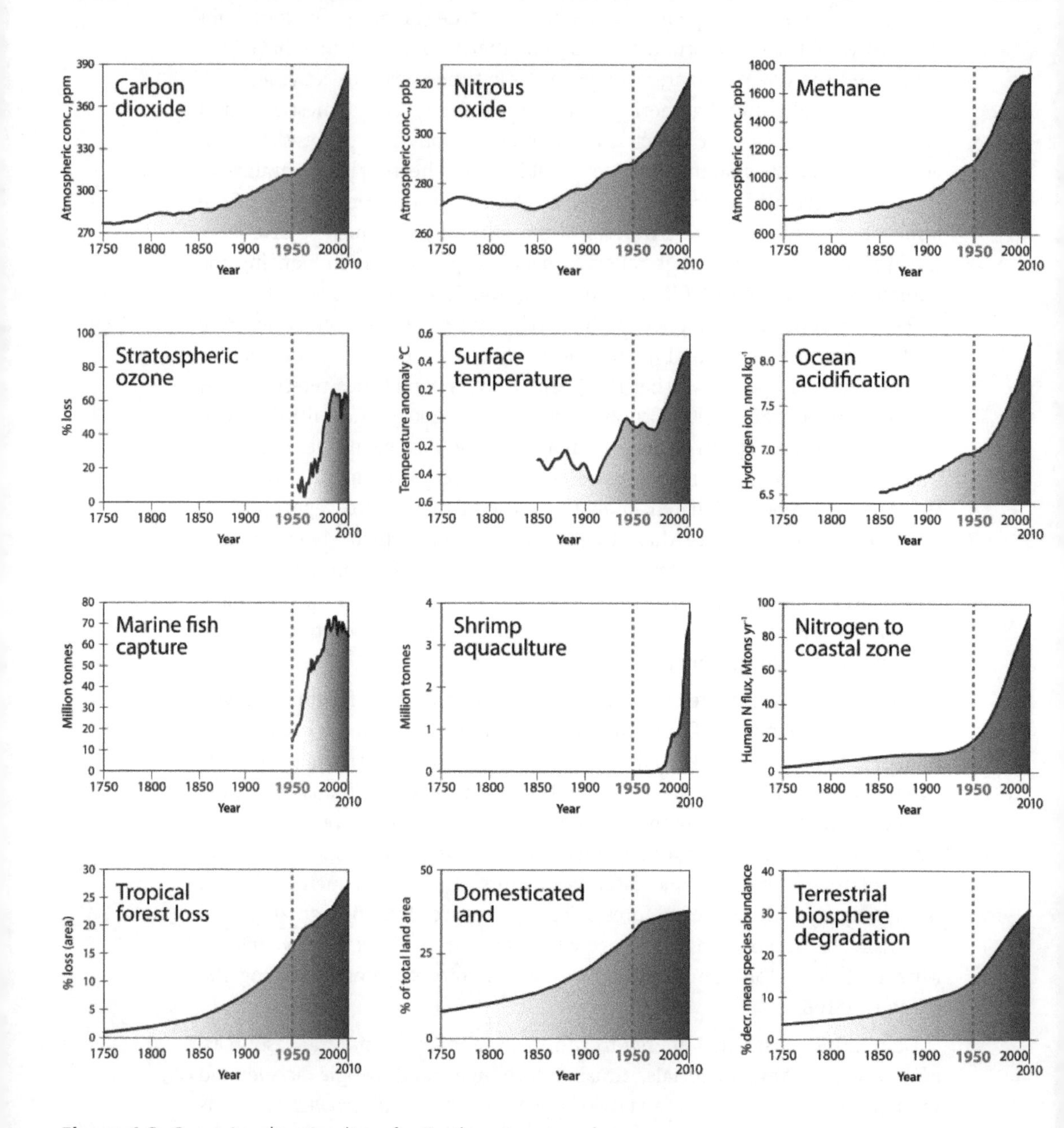

Figure 1.2 Great Acceleration lines for Earth system trends.
Source: IGBP 2015

Socioeconomic trends

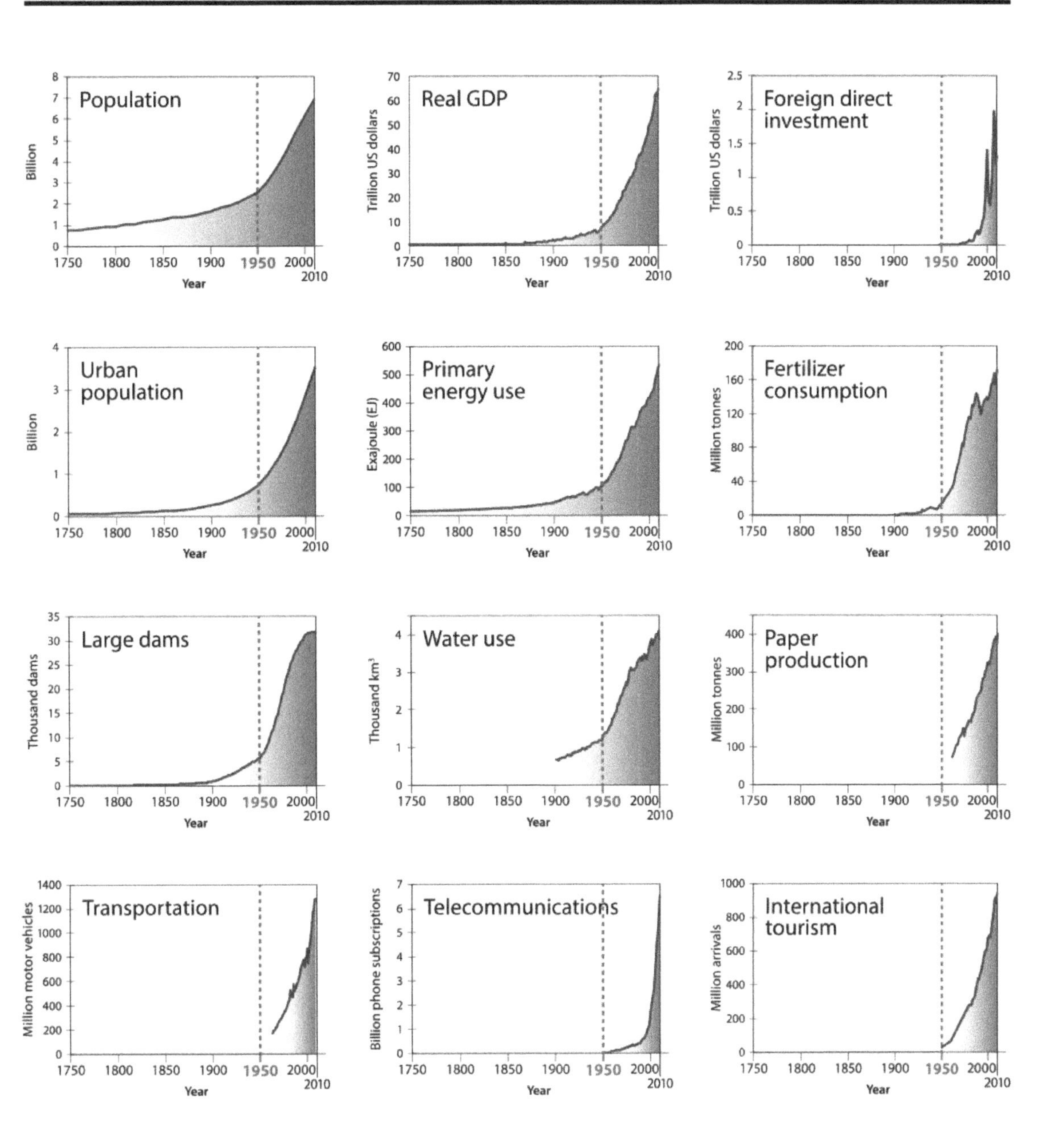

Figure 1.3 Great Acceleration lines for socioeconomic trends.
Source: IGBP 2015

to dramatically rising atmospheric concentrations of greenhouse gases, including carbon dioxide (CO_2), methane (CH_4), and nitrous oxides (NO_x). The Great Acceleration continues to the present day and is closely intertwined with societal changes, including population growth, economic globalization, urbanization, and rising levels of consumption. Human-induced climate change represents a key aspect of global environmental change, but it is not the only one associated with the Anthropocene. Humans are also contributing to biodiversity loss, changes in the geochemical cycles, land-use changes, ocean acidification, and many other impacts. One group of scientists has identified nine **planetary boundaries** to describe how human activities threaten to disrupt the Earth's physical systems and its "safe operating space for humanity" (Rockström et al. 2009; Steffen et al. 2015b) (see Table 1.1). By some accounts, humans have already overstepped four planetary boundaries: climate change, biosphere integrity, land-system change, and biogeochemical cycles (phosphorus and nitrogen) (Steffen et al. 2015b). Of particular concern are potential **tipping points** or thresholds where systems can shift irreversibly into another mode of operation, with unpredictable and potentially dangerous consequences. Examples of climate-related tipping points include Arctic summer sea ice, the Greenland ice sheet, the West Antarctic ice sheet, the Atlantic Thermohaline Circulation, Amazon rain forest dieback, the dieback of boreal forests, and shifts in the **El Niño–Southern Oscillation (ENSO)** (Lenton 2013).

Although the Anthropocene is a widely discussed concept within academic and policy circles, its definition, interpretation, and use are contested (Lövbrand et al. 2015; Dalby 2016). Many researchers are critical of the concept of the Anthropocene, pointing out that not every human being has had a

Table 1.1 Planetary boundaries

Biogeochemical flows	Nitrogen	Beyond zone of uncertainty (high risk)
	Phosphorus	Beyond zone of uncertainty (high risk)
Biosphere integrity	Genetic diversity	Beyond zone of uncertainty (high risk)
	Functional diversity*	Boundary not yet quantified
Land-system change		In zone of uncertainty (increasing risk)
Climate change		In zone of uncertainty (increasing risk)
Ocean acidification		Below boundary (safe)
Stratospheric ozone depletion		Below boundary (safe)
Freshwater use		Below boundary (safe)
Atmospheric aerosol loading*		Boundary not yet quantified
Novel entities*		Boundary not yet quantified

* *No global quantification. Source*: Based on Steffen et al. 2015b

negative impact on the planet (Biermann et al. 2016). Indeed, some argue that a relatively small part of the population has had disproportionately large consequences for the global environment (Palsson et al. 2013). Emphasizing global capitalism, unequal power relations, and materialist consumer culture as drivers of environmental degradation, some suggest that "**Capitalocene**" might be a more appropriate label for the present epoch (Altvater et al. 2016). Others draw attention to the role of patriarchy and gender inequalities as key processes that have shaped human–environmental interactions and suggest that dominant views of the Anthropocene offer restricted understandings of the entangled relations between natural, social, and cultural worlds (Gibson-Graham 2011). Some suggest that the urgent and apocalyptic message of the Anthropocene may be used to justify risky and undemocratic policy responses, such as geoengineering and totalitarian governance (Swyngedouw 2013; Lynch and Veland 2018).

The idea that we are "in" the Anthropocene coincides with a period of rapidly changing social, political, and cultural conditions. While more and more people recognize that humans are not separate from nature or

Box 1.2 Climate change over geological time

One way to gain perspective on human influence on planetary systems is to consider human history relative to geological time. Geological epochs mark major periods in the Earth's roughly 4.5 billion-year history (see Figure 1.4). To put humans into the context of geological time, the ancestors of humans have been around for about six million years, modern humans evolved about 200,000 years ago, and civilization as we know it only emerged about 6,000 years ago. The significant effects of human actions on the Earth system are evident in the atmosphere, biosphere, hydrosphere, and cryosphere. The pervasive signature of human actions is also reflected in the Earth's stratigraphy, which now includes deposits of aluminum, concrete, plastics and other human-made materials. The geochemical signatures in lake strata show high levels of petrochemical residues, pesticides, and other markers of human activities (Waters et al. 2016). This evidence has led scientists to propose the Anthropocene as a new geological epoch.

Consideration of geological time gives us a broader perspective on the magnitude and speed of recent and projected climate change. When we look at geologic time scales, especially prior to human history, we can see evidence of significant past climatic changes. For example, almost 56 million years ago, during a geological period referred to as the Paleocene–Eocene Thermal Maximum (PETM), global average temperatures increased by about 6°C (McInerney and Wing 2011). However, this increase occurred over a period of 20,000 years. Over the past 135 years, global average land and ocean surface temperatures have increased by nearly 1°C, and some projections suggest that temperatures may increase by an additional 2.6°C–4.8°C by 2100 (IPCC 2014a).

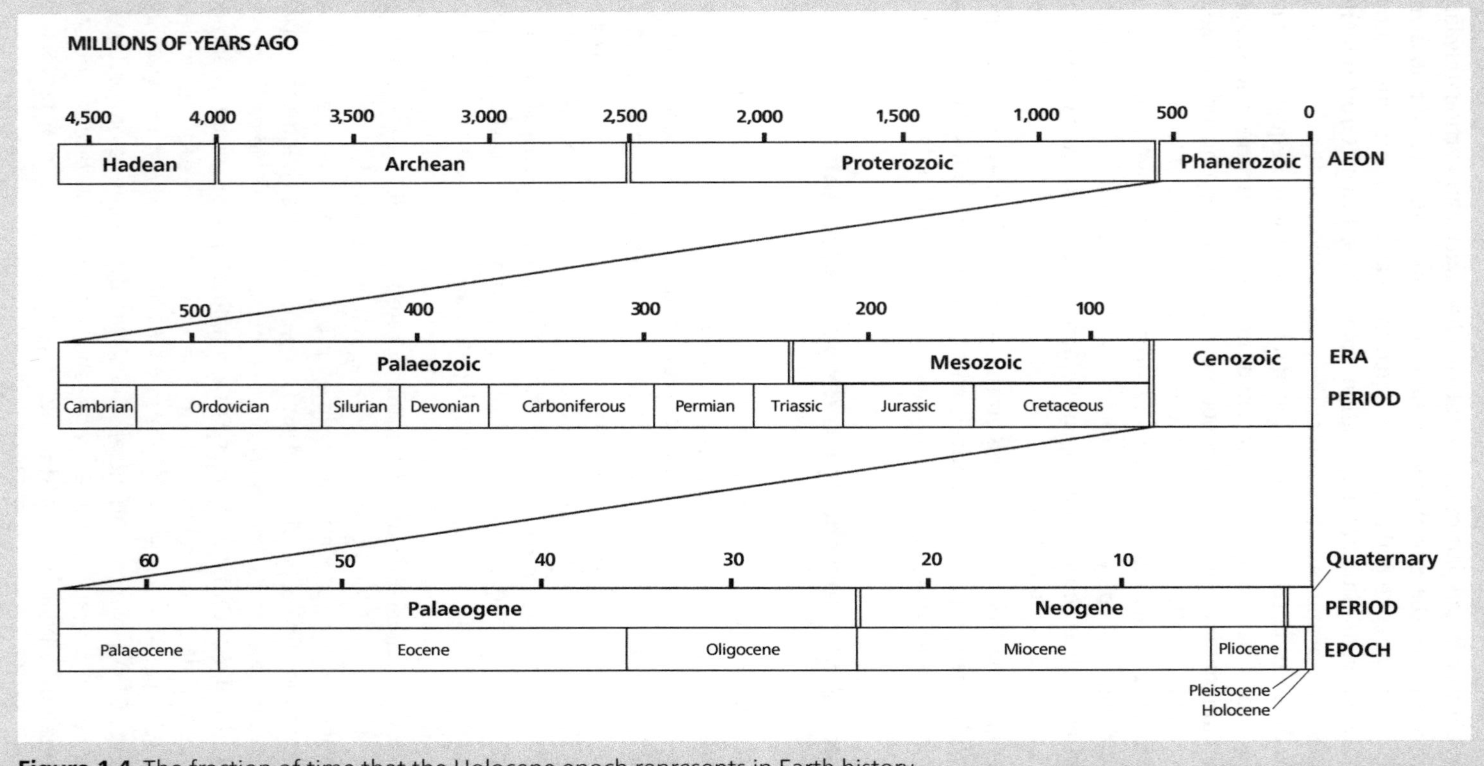

Figure 1.4 The fraction of time that the Holocene epoch represents in Earth history.
Source: Based on a figure from *Economist* 2011

each other and are advocating for social, economic, environmental, and **climate justice**, growing feelings of social isolation, economic insecurity, xenophobia, and nationalism have contributed to skepticism towards science, disdain for the future, and rejection of climate change as a valid concern. Some suggest that the Anthropocene can serve as a bridging concept that contributes to more inclusive understandings of complex issues such as climate change (Brondizio et al. 2016). Through its interpretive flexibility, the Anthropocene can potentially bring together research and insights from the natural sciences, social sciences, and humanities to create new narratives about humanity's relationship to the Earth, thereby expanding the range of futures that are considered possible and desirable (Castree 2014; Lövbrand et al. 2015).

The idea of the Anthropocene ultimately draws attention to human relationships with nature and to questions of ethics and responsibility for the future. Indeed, there have been increasing calls for a new ethics for the Anthropocene – one that includes responsibility and fairness at its core, embraces multiple histories, and recognizes that western ideas of mastery over nature are outdated (Schmidt, Brown, and Orr 2016). This new ethics entails inclusion of humans, non-humans, and future generations into everyday awareness and circles of care. Such an ethics suggests that humans living today have a moral duty to preserve the health of the planet for future generations.

Openings and opportunities

While much of our discussion so far has emphasized the risks associated with climate change, it also introduces openings and opportunities for transformational ideas, policies, and actions. Seeing climate change as an opportunity involves recognizing the threat and focusing on the possibilities and potentials for creating alternative futures, while at the same time reducing risks and vulnerabilities. As a shared danger, some view climate change as an opportunity for humanity to collaborate on solutions that benefit everyone. This view sees climate change as a mandate for global cooperation, with peace and unity as a potential outcome of the current crisis.

The signing of the **Paris Agreement** on climate change by more than 190 nations in 2015 might be seen as an example of the potential for global collaboration. The Paris Agreement is aimed at limiting warming to 2°C – and ideally to 1.5°C – by the end of this century (UNFCCC 2015). It is a potentially pivotal event towards realizing the goal of avoiding dangerous climate change, as set forth in Article 2 of the **United Nations Framework Convention on Climate Change** (UNFCCC). The specific UNFCCC goal is to stabilize emissions of greenhouse gases at a level that would "prevent dangerous anthropogenic interference with the climate system." While a precise threshold for what is dangerous has not been established, the general consensus is that temperature increases beyond 2°C would pose grave threats to human and ecological systems. It is important to remember, however, that even small changes represent dangerous climate change to some communities and groups, thus the difference between the 1.5°C and 2°C goals is significant (IPCC 2018b; Schleussner et al. 2016). Although the commitments are voluntary, the 1.5°C target represents more than a temperature goal. Rather, it can be interpreted as a shared recognition that the implications of climate change are serious, unevenly distributed, and costly, and that the burdens of adapting to climate change will increase as global temperatures rise (IPCC 2018b). In many ways, the Paris Agreement can be thought of as "marching orders" for a global transformation to sustainability.

Framing climate change as an opportunity can also spark new research and innovation, for example on renewable energy technologies, early warning systems, or changes in food access or consumption practices. Responding to climate change will create many new types of jobs and foster new industries aimed at the production and distribution of alternative forms of energy and materials, new learning platforms, and new modes of transportation and social interaction. Climate change can also catalyze and expand movements that link environmental and social justice, challenging exclusionary and oppressive norms and practices that are contributing to risks and vulnerabilities. These movements have a diversity of goals, from increasing awareness of the unequal impacts of climate hazards to adopting more just and sustainable modes of living and caring for the Earth. A growing number of governments, NGOs, businesses, civil society groups, and citizens recognize that addressing climate change can promote innovative forms of governance, including more equitable and sustainable ways to organize the economy and society.

Climate change also presents an opportunity for deeper inquiry into how mindsets, beliefs, values, and worldviews influence **social norms**, institutions, and economic relationships (O'Brien 2018). There are growing questions about whether narratives of "progress," which have contributed to industrialization, economic growth, and globalization, are suited to the challenges of the twenty-first century (Prudham 2009). Alternative narratives are emerging (or being revitalized), along with new **paradigms** that emphasize collaboration, connection, and integration over competition, separation, and fragmentation (Veland et al. 2018). Also relevant is the idea of social tipping points, where small changes lead to abrupt shifts in social norms and values that may transform larger social systems towards sustainability (Bentley et al. 2014). Artists, intellectuals, and members of civil society are forming communities of practice to promote deeper cultural changes that are visible in everything from fashion to literature to education. A recognition of animal rights, the value of ecosystems, and the connections between humans and the environment are contributing to a revision of the atomistic, mechanistic, and deterministic understandings of nature that emerged during the Age of Enlightenment. For example, some argue that we are moving into the "Age of Enlivenment" where consciousness is seen to be a vital part of all life (Weber 2013).

The barriers to limiting global warming to 1.5°C or 2°C remain formidable. In fact, some argue that avoiding dangerous climate change is now impossible, and that we simply need to adapt to the impacts of climate change and accept the losses. Others point out that there are biophysical and social limits to adaptation as a response to climate change, and that successful adaptation will require social transformations (O'Brien 2017). It is important to be clear about the many different assumptions regarding the possibilities and potentials for societal change and to challenge them when necessary. It is also important to point out that not *everything* needs to be transformed. In fact, many existing institutions, values, norms, and ways of living and working are already contributing to equitable climate change responses (Bennett et al. 2016). Finally, it is important to acknowledge the historical significance of humanity's capacity to innovate and collaborate, especially in times of crisis (Butzer and Endfield 2012). The potential to mobilize and engage society with transformations to sustainability is perhaps the biggest opportunity associated with climate change.

Climate change as transformation

Climate change demonstrates that humans are capable of transforming global systems. Through our activities, we are influencing the climate system in fundamental ways – changing temperatures, rainfall patterns, extreme events, and sea levels. We are also contributing to both risks and vulnerabilities through the ways that we organize society, including policies and regulations, institutions, incentives, and social practices. In many respects, climate change is a wake-up call that encourages a questioning of assumptions about power and politics, economic growth, resource-based consumption, material prosperity, and **human exceptionalism**. The types of transformations that we are causing through climate change are not random or inevitable but are a reflection of how we think about ourselves in relation to the natural world, in relation to each other, and in relation to the future.

As we emphasize throughout this book, there are many ways to frame, understand, and explore climate change and its relationship to society. Geographer Mike Hulme (2009) refers to the "plasticity of climate change" to describe how the idea of climate change takes on different meanings in different contexts. Each framing of climate change sees the problem in a different way and prioritizes different kinds of solutions. Some approaches focus on specific scientific or policy dimensions of the issue, while others connect climate change to large-scale, social processes, particularly globalization, urbanization, and the expansion of a consumption-oriented economy. Holistic approaches are beginning to recognize the emotional, psychological, and spiritual dimensions of climate change and link them to political actions and interventions that support sustainable development (Kiehl 2016).

This book presents an integrative framing of climate change, exploring the many lenses through which one can view the issue and its social dimensions. The overall goal of the book is to help readers think about climate change in new and different ways, to see not only the problems but also the possibilities and potentials for transformative responses. Below, we highlight some of the key messages that will emerge as you read this book:

1 *Science is not enough to stimulate action on climate change.* As we discuss in chapter 2, climate science provides overwhelming evidence that humans are causing climate change. Yet scientific knowledge about climate change is not sufficient to engage society with meaningful responses. Research from the social sciences and humanities draws attention to the role of power relationships and vested interests in maintaining the status quo and suggests that a lack of facts and information is not the problem. Action on climate change is influenced by whether and how the information resonates with values, identities, and emotions, and how it penetrates the social and political barriers to change. It also points to the power of narratives, stories, and dialogues to engage people personally and politically with climate change.

2 *Discourses have power.* Different ways of defining, describing, and discussing the problem of climate change lead to different solutions. In chapter 3, we present some of the key discourses related to climate change and describe the types of responses that they prioritize. We emphasize that responses are not neutral but instead are closely linked to ideas and understandings of what knowledge is and whose knowledge is legitimate. Discourses are internalized and reproduced through media, education, and other social institutions, and they are often actively promoted by those with interests in maintaining or achieving a particular outcome. Recognizing the power of language and ideas to shape society can also allow us to identify discourses with the potential to transform society.

3 *Worldviews shape our perspectives.* Worldviews, along with the values, beliefs, and assumptions that support them, influence how we relate to and approach climate change. In chapter 4, we explore how worldviews represent the lenses through which we understand reality; like discourses, they are shaped by social processes, institutions, and relations, as well as by everyday

experiences. The seeming intractability of climate change presents an opportunity to widen the lens and to view problems and solutions from new and different perspectives. This introduces a significant role for art and imagination in visioning alternative futures.

4 *Greenhouse gas emissions and energy use are driven by social processes.* In chapters 5 and 6, we show that the rising emissions from the production and consumption of fossil fuels are intertwined with long-term processes of development, globalization, and urbanization. Addressing the drivers of climate change through mitigation policies and practices, such as promoting renewable energy and sustainable consumption patterns, will also entail challenging systems, power relations, and vested interests, as well as social norms, values, and mindsets.

5 *Climate change impacts are highly unequal but affect everyone.* Chapters 7 and 8 demonstrate that climate change is having profound yet highly uneven impacts on individuals, households, communities, nations, and cultures, as well as on other species and ecosystems. Differences in where people live and how they earn their living, as well as disparities in health, assets, and resources, influence how climate change will be experienced, as well as the ability to respond. Yet in a highly interconnected and globalized world, everyone's security will be directly or indirectly affected by climate change.

6 *Adaptation to climate change is necessary but insufficient.* In chapter 9, we turn to responses to climate change, focusing on adaptation. While responses that mitigate future climate change are vitally important, some changes are unavoidable because CO_2 emissions from past activities will continue to warm the planet for many decades. As a result, adaptation is necessary to prepare society to live with the impacts of climate change. Yet losses will be experienced, and there are both barriers and limits to adaptation. Ultimately, adaptation to climate change offers opportunities for humans to transform their relationships with the environment, with each other, and with the future.

7 *Transformations to sustainability are possible and underway.* In chapter 10, we emphasize that the future we experience will, to a large extent, depend on human choices, decisions, and actions. We present an integrative lens on transformative change that emphasizes connections among the practical, political, and personal spheres of transformation. Exploring various entry points for transformation, we emphasize the role of individual and collective agency in promoting a sustainable, equitable, and thriving world.

Summary

Many questions remain about whether and how society can make the profound changes needed to avoid dangerous climate change, especially when it remains an abstract or irrelevant issue to many people, when other concerns are considered priorities, and when people with vested interests and power are committed to the status quo, regardless of the implications for others, including future generations. As we emphasize throughout this book, climate change is not simply a physical phenomenon that affects human systems. It is a social issue that reflects constellations of political and economic power, cultural norms, and expectations about prosperity, well-being, quality of life, and security. Climate change is about how we see ourselves in relation to nature and to others and how we organize ourselves and our societies. It is a global challenge that requires acknowledging the diverse ways that change is understood and addressed. Meeting the challenge of climate change can open new opportunities and possibilities for a better world.

Reflection questions

1. Do you think that climate change matters for your life? What aspects of climate change are most important to you, and why?
2. How and why are issues other than climate change relevant to the Anthropocene?
3. In what ways do you think that climate change might be an opportunity? For whom?

Further reading

Dalby, S., 2016. Framing the Anthropocene: The good, the bad and the ugly. *The Anthropocene Review* 3(1): 33–51.

Gibson-Graham, J. K., 2011. A feminist project of belonging for the Anthropocene. *Gender, Place & Culture* 18(1): 1–21.

Hulme, M., 2009. *Why We Disagree about Climate Change: Understanding Controversy, Inaction and Opportunity*, 4th edn. Cambridge, UK: Cambridge University Press.

O'Neill, B. C., Oppenheimer, M., Warren, R., et al., 2017. IPCC reasons for concern regarding climate change risks. *Nature Climate Change* 7(1): 28–37.

Steffen, W., Richardson, K., Rockström, J., et al., 2015b. Planetary boundaries: Guiding human development on a changing planet. *Science* 347(6223): 1259855-2–1259855-10.

2 SCIENTIFIC EVIDENCE OF CLIMATE CHANGE

The scientific consensus

We often hear that there is a scientific consensus on climate change – in other words, most of the world's climate scientists agree that humans are having an impact on the global climate system (Cook et al. 2016). Scientific findings that link human activities to the climate system have played a decisive role in identifying the problem of climate change and in justifying action to address it. Yet many people still have questions about the science of climate change: Hasn't the climate always changed? How do we really know that human activities are causing climate change? Why would a few degrees of warming matter? Will more and better scientific information on climate change lead to action?

This chapter presents the dominant scientific narrative on climate change. Rather than attempting to provide a comprehensive treatment of this broad topic, we focus on the physical and statistical evidence of climate change and highlight issues and questions that are especially relevant to climate and society relationships. We begin by exploring evidence that the climate is changing and then discuss the connections between climate change, greenhouse gas emissions, and human activities. We also consider projections of future climate change based on climate models and the geological record. After reviewing the scientific evidence, we turn to the question of whether science can motivate action on climate change. We discuss research on climate change communication, which explains why a scientific narrative does not necessarily resonate with all audiences. We conclude by underscoring that scientific evidence provides strong justification for action to reduce emissions of greenhouse gases, but that science alone is insufficient to catalyze large-scale societal responses.

Evidence of climate change

There is evidence all around us that the climate is changing. Winters are getting warmer, trees and flowers are blooming earlier in the spring, summer heat waves are becoming longer and more severe, glaciers are melting, bird migration patterns are changing, and tropical diseases are becoming more prevalent at higher latitudes. In this section, we explore different forms of scientific evidence of climate change.

Documenting a changing climate

Scientific evidence of a changing climate has been accumulating rapidly in recent decades. Rising global temperatures are perhaps the most widely documented indicator of climate change. Global mean temperatures have increased by almost 1°C (1.4°F) over the past 135 years (IPCC 2018b). The majority of this warming has happened in recent decades. With the exception of 1998, 17 of the 18 warmest years over the 136-year record of global surface temperatures have occurred since 2001 (NASA 2018). The period from 2014 to 2016 marked the first time that the global temperature record was exceeded in three consecutive years, with each year breaking the record set in the previous year (NASA 2017).

Data and graphics on global average temperature changes over time can give the impression that temperatures are rising at the same rate all over the planet (Figure 2.1). A closer look at regional patterns, however, reveals that the distribution of temperature change is spatially uneven. In other words, a 1°C increase in global average temperatures entails larger temperature increases in some places and smaller changes, or even cooling, in other places. Maps showing how temperatures around the globe have deviated from averages provide one way of identifying locations where temperature changes have been greater or less than the global mean (Figure 2.2). High-latitude areas, including the Arctic, have experienced particularly large temperature increases over recent decades – a warming that has to some extent been offset in other regions by the cooling effect of aerosols in the atmosphere (Najafi, Zwiers, and Gillett 2015). The signal of warming temperatures is more apparent in high-latitude regions, where weather is less sensitive to shifts in the jet stream, the El Niño–Southern Oscillation (ENSO), and other forces that contribute to increased variability in the mid-latitudes.

Evidence of climate change can also be observed by looking at trends in **climate normals**, which are defined as the average of weather in a region over a 30-year period (see Box 2.1). Climate normals are typically updated every ten years, and a comparison of values across regions and over time can reveal long-term patterns and trends while filtering out short-term fluctuations or anomalous years. Other types of temperature measures also serve as indicators of climate change. For example, trends in daily maximum and minimum temperatures can reveal how climate change is affecting the daily temperature range. In many regions, both daily maxima (representing

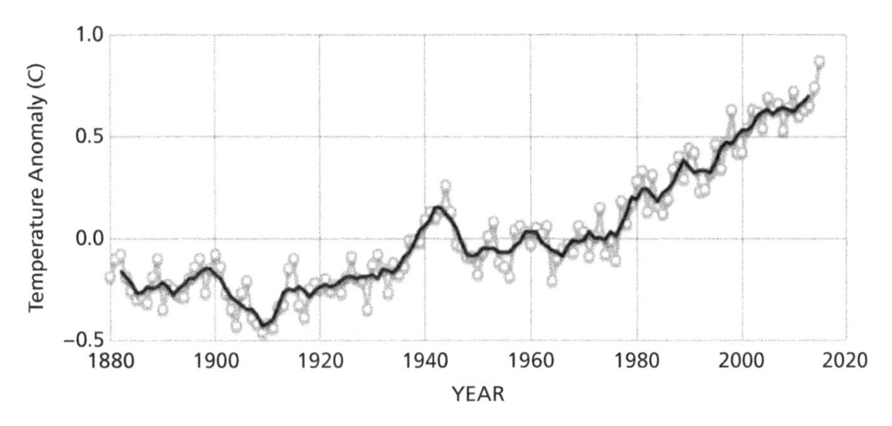

Figure 2.1 Global mean temperatures over time.
Source: NASA/GISS 2018

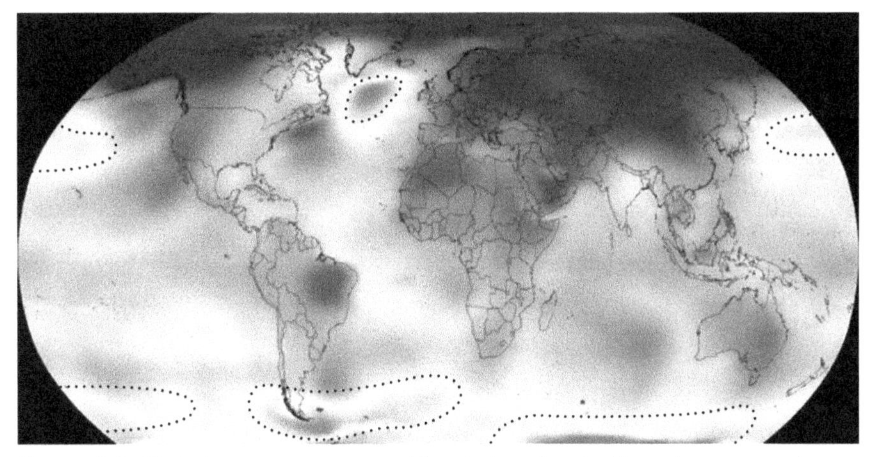

Figure 2.2 Global temperature map. The darker the shading, the greater the increase in temperature, except for the marked regions where temperatures are decreasing.
Source: NASA/GISS 2018.
For full view, please visit: https://climate.nasa.gov/vital-signs/global-temperature

the warmest temperatures in a 24-hour period, typically experienced in the afternoon) and minima (representing the coldest temperatures in a 24-hour period, often experienced at night) are higher than they used to be. Changes in average temperatures during particular seasons (e.g., hotter summers or warmer winters), as well as changes in the timing of particular seasonal events (e.g., first frosts or last snowfall), are also evident, especially in mid- and high-latitude regions. As we discuss in chapters 7 and 8, changes in the daily and seasonal distribution of temperatures have a wide range of implications, for example on crop yields, human health, and extreme events such as wildfires.

In addition to changes in temperature, there is also evidence of significant changes in precipitation patterns around the world (Trenberth 2011). Observed

Box 2.1 Weather versus climate

Weather is one of the most talked about phenomena in the world. An internet search on "weather" returns more than 1.2 billion results – and this is only for the English word. Weather apps are some of the most frequently used smartphone features (Phan et al. 2018); weather segments of news shows and dedicated weather programs are also widely watched. Yet it is common for people to mix up weather and climate, leading to confusion about what climate change actually means. A significant snowfall in the middle of winter may be mistakenly interpreted by some as evidence that the climate is not changing, whereas an unseasonably warm winter day could be taken by some as a signal of climate change. In both cases, it is important to look at the trends and patterns of variability over time. Comparison of recent annual temperatures with the climate normal from an earlier baseline period (such as 1960–1990) can help to illustrate how annual temperatures have changed over time (see Figure 2.3).

Weather describes the outdoor conditions at a particular time or on a particular day in a specific location, whereas climate is a statement about average weather conditions over several decades in a location or region. For example, the January climate in New York City is characterized by temperatures around 0°C (or 32° in °F) and precipitation in the form of snow. However, the weather on a particular

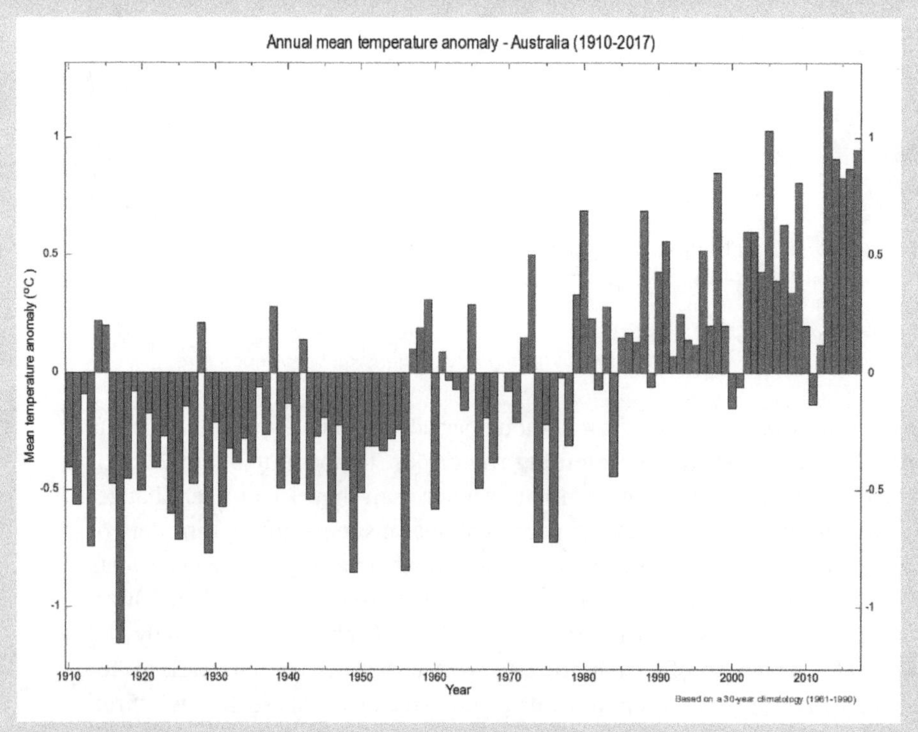

Figure 2.3 Temperature deviations from normals for Australia.
Source: Australian Bureau of Meteorology

day in January in New York might be in the middle teens (°C) (or the sixties in °F) and rainy. Weather can be highly variable, sometimes changing several times during a single day, whereas climate is relatively stable. In Melbourne, Australia, for example, the weather on any given day can change dramatically between morning, noon, and evening, such that one can experience "four seasons in one day." Nonetheless, the climate in Melbourne is still characterized as temperate, with warm summers and cool winters. In many locations around the world, it is often said that "climate is what you expect; weather is what you get."

Do you think that confusion about the difference between weather and climate plays a key role in public misunderstandings of climate change? What steps, if any, do you think should be taken to improve understanding of these scientific concepts?

precipitation patterns vary, but there is a general trend showing that dry areas are experiencing drier conditions, and wet areas are experiencing more rainfall, often in the form of more extreme precipitation events, such as intense downpours (Fischer and Knutti 2015) (see Box 2.2). That being said, there are many exceptions to these patterns. In recent decades, the Amazon rain forest has experienced unprecedented dry spells, introducing water scarcity to communities and ecosystems that are accustomed to abundant rainfall (Marengo and Espinoza 2016).

Observed changes in precipitation are a result of several interacting factors (Trenberth 2011). First, warmer temperatures lead to more evaporation,

Box 2.2 Extreme weather and climate change

While it is difficult to identify the precise effect of climate change on any single extreme event, climate change "loads the dice" to make extreme events more likely to occur than would otherwise be the case. To understand this better, we can use a pair of dice as an example. The effect of climate change on the likelihood of an extreme event is frequently compared to the effect of rolling a pair of loaded dice (Hansen, Sato, and Ruedy 2012). When you roll a regular pair of dice, there is a 1/36 (1/6 times 1/6) chance that you will roll double sixes. When you roll a pair that is loaded towards 6 (i.e., the dice are each weighted to make rolling six more likely), then your chances of rolling double sixes are much higher. You will not always roll a double 6 when the dice are loaded, but your odds of rolling them are certainly higher than when you roll a pair of regular dice. By adding greenhouse gases into the atmosphere, we have been loading the dice in favor of more variable and extreme weather.

which makes more water available for precipitation. Second, a warmer atmosphere holds more water vapor, which can lead to heavier downpours. Third, changes in large-scale atmospheric circulation patterns influence the transport of moist air over oceans and land. Together, these factors influence where, when, and how much rain or snow falls. While changes in average precipitation may not be substantial, shifts in the timing and intensity of rainfall are often highly disruptive, contributing to both floods and droughts. For example, the city of Houston received 1,318 mm (51.88 inches) of rain during Hurricane Harvey in 2017, setting a new record for precipitation from one storm in the continental United States (Blake and Zelinsky 2018). By contrast, in the city of Cape Town, South Africa, rainfall has been below normal for several years in a row, leading to a severe water crisis in 2018, which resulted in a rationing of supplies for most of the city's residents (Maxmen 2018).

Although we have been focusing on changes related to the atmosphere, changes in ocean temperatures and chemistry have also been observed. The atmosphere and oceans can be thought of as two fluids that are constantly interacting. Rising temperatures heat water molecules in the ocean, and warmer water occupies a greater volume than cold water. Most (90%) of the warming of the atmosphere over the past two centuries has been absorbed by oceans, and the thermal expansion of water is contributing to higher sea levels worldwide. Warmer air temperatures are also melting glaciers and ice sheets on land, leading to freshwater runoff that adds to the volume of water in oceans. Measurements of average global sea levels show an increase of about 20 cm (roughly 8 inches) since the middle of the nineteenth century (IPCC 2014a). About half of this can be attributed to the warming of the oceans and half to glacial melting. As we discuss in chapter 7, sea-level rise poses a threat to low-lying coastal areas, including many Small Island Developing States (SIDS), not only because of higher average sea levels but because of both higher storm surges and increased coastal erosion that result when winds or tides push a larger volume of water onto the shore.

Other evidence of climate change

Evidence of climate change comes from many sources. Temperature and precipitation data are collected through a global network of measurement stations located on land and oceans and from satellite data. Other evidence includes tree-ring measurements from forests in Northern California, harbor records that document historical ice melting in the port of Tallinn, Estonia, wine records from Burgundy, France that show trends in harvest dates, or comparisons of photographs of Andean glaciers taken early in the twentieth century with photos taken today. Although such alternative measurements do not directly capture climatic conditions, they provide valuable information about past climates and the effects of human activities, including deforestation and urbanization. In Japan, for example, phenological data extending back for hundreds of years are used to track the flowering times of cherry trees. An analysis of these data shows that the trees are blooming earlier than ever. While climate change is a key culprit, local urban heat island effects are also contributing to earlier blooming (Primack, Higuchi, and Miller-Rushing 2009).

Evidence of climate change also comes from observations made by individuals, including hunters, farmers, gardeners, bird-watchers, and others whose work, hobbies, or daily lives are directly affected by changes in weather and climate. Inuit hunters in Alaska and Canada are experiencing changing ice conditions that are affecting their ability to safely hunt seals and other animals (Hovelsrud and Smit 2010). Farmers and home gardeners in temperate areas are finding that the crops and flowers they have always planted no longer thrive because the climate has become hotter or drier or because pests and diseases that are new to the region are causing damage. Bird-watchers are discovering that migratory species such as the American robin, which used to fly south for the winter months, are staying in northerly regions for much of the year (Wood and Kellermann 2015). In coastal communities from Florida and Virginia to the Marshall Islands, flooded property and impassible roads are a regular occurrence during bi-monthly high tides (Sweet et al. 2014). Shifting seasonal patterns and a longer monsoon in Bangladesh are reducing the number of distinguishable seasons from six to four (Islam and Kotani 2016). In Greenland, villagers are noticing more and more tourists coming to see the melting glaciers. At the same time, the disappearance of ice is making it difficult for them to fish and hunt in the winter (Hastrup 2016).

It is not just the average climate conditions that are changing. There is also evidence of changes in climate variability and in the frequency and magnitude of extreme weather events (Hansen, Sato, and Ruedy 2012). Record-shattering temperatures, rainfalls, winds, storms, wildfires, and floods reflect a dynamically changing climate system. There are numerous reasons to expect that climate change is contributing to increases in the frequency of extreme events. For example, changes in physical processes, such as warmer air and ocean temperatures, influence the intensity of rainfall and heat waves (Fischer and Knutti 2015). Extended periods with dry conditions can lead to more wildfires, and changes in the path of the jet stream can lead to record cold spells (NAS 2016).

The question of whether or not extreme weather events, such as heat waves and storms, are linked to climate change comes up frequently in media coverage of weather events. Casual observations, such as "it's never been this hot in May," or "we didn't used to have such heavy rains," have become commonplace in many communities. In the United Kingdom, for example, unpredictable weather has many people talking about "global weirding." The scientific community is cautious about linking extreme weather events directly to climate change. The challenge is to separate out short-term climate variability from long-term trends and to show that extreme weather events are statistically significant. Identifying the **climate change fingerprint** of an extreme event provides strong support for attribution of impacts to climate change. Yet even in the absence of statistical evidence linking a particular extreme event to climate change, the expectation of more frequent and severe extremes is consistent with climate change (see Box 2.2) (NAS 2016).

Taken as a whole, evidence of rising temperatures, changing rainfall patterns, warming oceans, melting glaciers, rising sea levels, and changes in climate variability and extreme events all point to changes in the global climate (see Figure 2.4). What appears to be a "normal" climate to young people today is very different from what their parents, grandparents, and great-grandparents understood and experienced. In fact, the scientific concept of a 30-year climate normal may no longer be an appropriate way to describe climatic conditions (Livezey et al. 2007). New patterns of increasingly warm years are starting to influence climate on decadal scales. The saying that "the future is not what it used to be" takes on new significance in a changing climate.

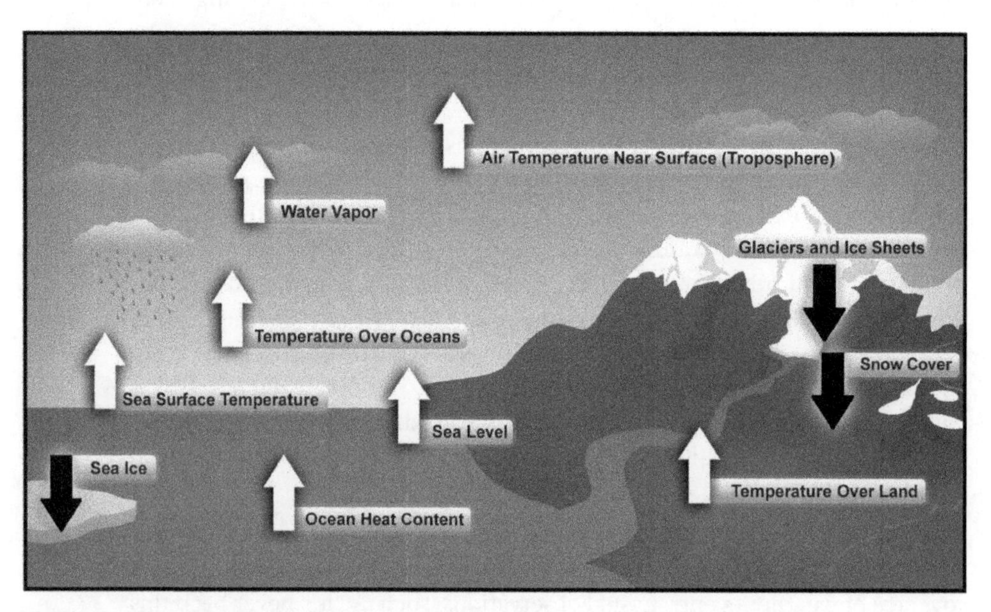

Figure 2.4 Ten indicators of a warming world.
Source: NOAA NCDC based on data updated from Kennedy et al. 2010

Linking climate change to greenhouse gas emissions and human activities

Scientific documentation of a changing climate is clear. Science also provides strong evidence that observed warming is the result of increasing emissions of greenhouse gases from human activities. An important pillar of this understanding of anthropogenic climate change is the earth's **radiative balance** – the relationship between incoming solar radiation and outgoing terrestrial energy (see Box 2.3). The radiation balance is influenced by the Earth's surface and the composition of the atmosphere, as well as by variations in cloud cover. Over longer time periods, it can also be influenced by changes in solar radiation, both due to sunspot cycles and changes in the Earth's orbit around the sun.

The greenhouse effect is a common way to think about the Earth's radiative balance (see Figure 2.5). The metaphor describes how the sun's energy passes through the Earth's atmosphere (i.e., the "glass" in the greenhouse) to warm the surface of the planet, including the land and oceans. The Earth's atmosphere consists of nitrogen (78%) and oxygen (21%), argon (0.9%) and very small amounts of other gases, such as water vapor, carbon dioxide (CO_2), methane (CH_4), ozone (O_3), and nitrous oxides (N_2O). These trace gases are referred to as greenhouse gases because they play a key role in retaining heat and keeping the planet warmer than it would otherwise be. The Earth's warm surface emits infrared thermal radiation back to space, with some of it absorbed by greenhouse gases in the atmosphere and reradiated in all directions, including back towards the Earth's surface. The warming effect of greenhouse gases helps to keep the Earth's temperature in a range that can support plants, animals, and human life. In fact, without greenhouse gases in the atmosphere, the Earth would be 33°C colder.

Awareness of the importance of the composition of the atmosphere for the Earth's radiative balance dates back to the work of Jean-Baptiste Joseph Fourier in the 1820s. His research recognized the warming effect of trace gases in the atmosphere. Later research by Eunice Foote and John Tyndall confirmed the greenhouse effect, and, towards the end of the 1880s, a more decisive link was established between changes in CO_2 and global temperature through the work of Swedish chemist Svante Arrhenius. Arrhenius calculated that rising levels of CO_2 as the result of the burning of fossil fuels would lead to an enhanced greenhouse effect that would change the Earth's radiative balance and result in the warming of the planet. While the physical basis for the greenhouse effect was firmly established by the end of the nineteenth century, it took until the middle of the twentieth century for greenhouse gas concentrations in the atmosphere to be methodically tracked. The well-known Keeling Curve (named after Charles Keeling who

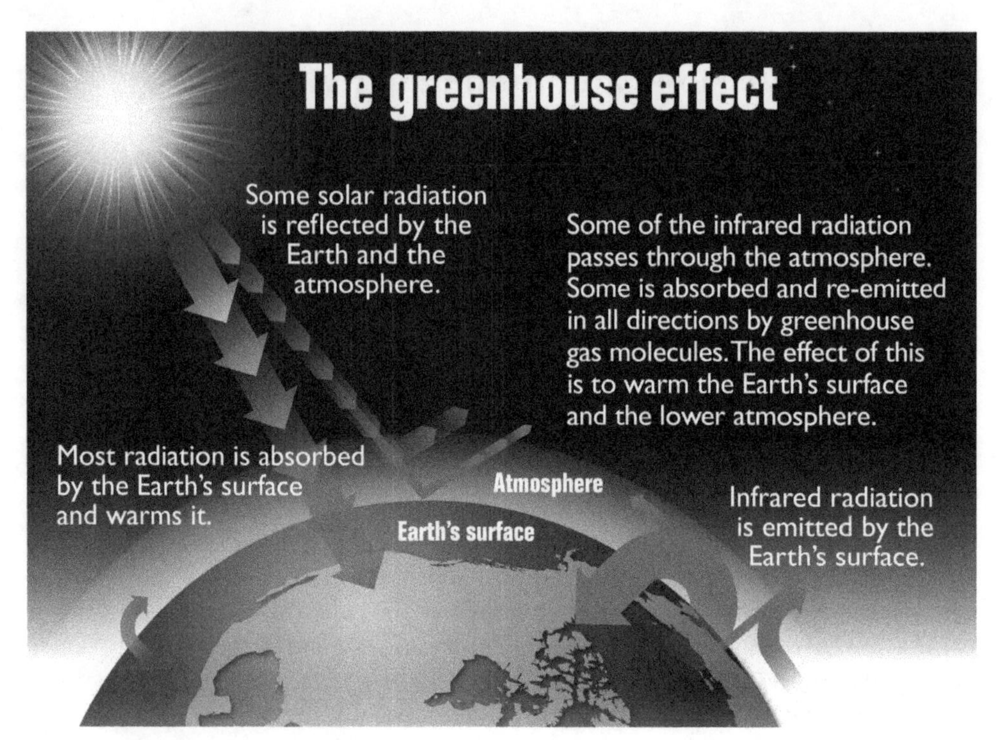

Figure 2.5 The greenhouse effect.
Source: US EPA 2012

began taking measurements of atmospheric CO_2 levels in Mauna Loa, Hawaii, in 1958) shows a steady rise in CO_2 concentrations over the past 60 years (see Figure 2.6). The first measurements from Mauna Loa, Hawaii, in 1958 showed atmospheric CO_2 concentrations of 316 parts per million by volume (ppmv, for short). Atmospheric CO_2 concentrations surpassed 410 (ppmv) for the first time in 2018 (NOAA 2018).

What is important in the Keeling Curve is the unmistakable upward trend. Not only have concentrations of atmospheric CO_2 been increasing over time, but the rate of increase has also been accelerating, especially over the past several decades. In examining the Keeling Curve, it is useful to note that the levels of CO_2 show seasonal variations related to the presence or absence of leaf cover, especially in the Northern Hemisphere, which contains 67% of the Earth's land area. During the Northern Hemisphere summer, vegetation takes up carbon as part of the process of photosynthesis and lowers atmospheric CO_2 concentrations; in the Northern Hemisphere winter, the absence of leaf cover corresponds with higher CO_2 concentrations. These annual variations are evidence of interactions between the atmosphere and the biosphere. Increasing CO_2 emissions are also absorbed by oceans and contribute to ocean acidification. In fact, oceans take up nearly a third of the carbon added to the atmosphere by humans, which is reducing pH and

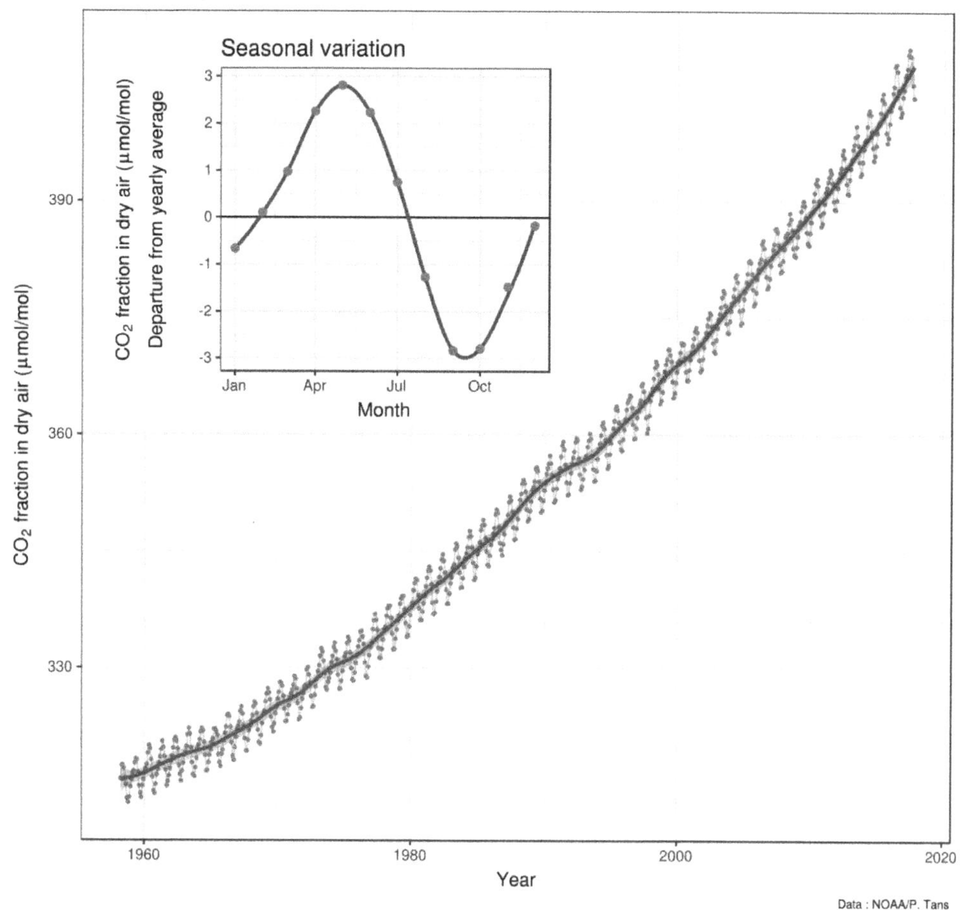

Figure 2.6 The Keeling Curve.
Source: Wikimedia commons. Data from Dr Pieter Tans, NOAA/ESRL and Dr Ralph Keeling, Scripps Institution of Oceanography.

altering chemical balances that are critical to marine ecosystems (Doney et al. 2009).

Increasing atmospheric concentrations of CO_2 have a significant influence on the climate system, and there is evidence of an almost linear relationship between cumulative CO_2 emissions and global temperature change (IPCC 2014a). Along with rising levels of CO_2, emissions of other major greenhouse gases, such as methane, nitrous oxide, and hydrofluorocarbons, have also been increasing since the middle of the twentieth century. Increases in these other gases are associated with human activities including agriculture, fossil fuel production and consumption, industrial processes, air conditioning, and refrigeration. Although the total quantity of these other gases in the atmosphere is far less than the quantity of CO_2, they each have substantially greater **global warming potential** than CO_2 and can therefore make significant contributions to an enhanced greenhouse effect.

The influence of human activities on the climate system is complex, and there are many uncertainties. The Intergovernmental Panel on Climate Change (IPCC) uses a calibrated language for evaluating and communicating the degree of certainty and confidence in particular findings about climate change (Mastrandrea et al. 2011). This is based on the type, amount, quality, and consistency of the evidence, as well as the agreement among different studies. With respect to warming temperatures and greenhouse gases, the IPCC (2013a: 17) states that "It is extremely likely that more than half of the observed increase in global average surface temperature from 1951 to 2010 was caused by the anthropogenic increase in greenhouse gas concentrations and other **anthropogenic forcings** together." This conclusion is based on scientific research and understanding of the climate system.

The evidence that humans are contributing to climate change is further supported by research that considers other possible explanations for the observed changes. Over very long time periods, changes in the shape of the Earth's trajectory around the sun, the tilt of its axis, and its "wobble" around this axis – referred to as the Milankovitch cycles – together influence the amount of solar radiation that the Earth receives. These factors, which vary over periods of 26,000, 44,000, and 110,000 years, have been linked to the onset of glacial and interglacial periods. Over shorter time periods, sunspots, volcanic eruptions, and smoke from forest fires can influence the climate. Short-term climate variability can also be traced to the El Niño–Southern Oscillation, which is the result of variations in sea-surface temperatures in the eastern Pacific Ocean. While all these factors play some role in climate trends, none can account for the rapid increases in temperature and other climate changes that have been observed since the early nineteenth century (IPCC 2014a). Evidence points to the rising concentrations of greenhouse gases resulting from human activities. These activities, including the burning of fossil fuels, as well as clearance of forests and other land-use changes, have been the major contributors to rising emissions and observed temperature changes (see Figure 2.7).

We will explore human drivers of increasing greenhouse gas emissions and the important roles of fossil fuels and land use in climate change in more detail in chapters 5 and 6. The point to keep in mind here is that human-induced increases in atmospheric concentrations of greenhouse gases are changing the Earth's energy balance. Energy that would otherwise escape to space is trapped in the atmosphere by these additional gases. As a result, the planet is warming.

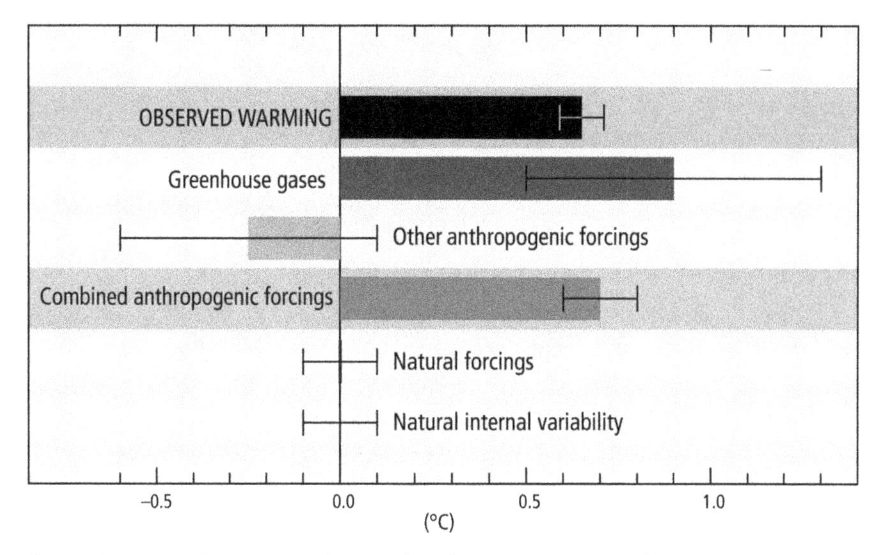

Figure 2.7 Contributions to observed surface temperature changes, 1951–2010.
Source: IPCC 2014a

Projecting climate futures

While it is clear that global warming is already underway, the precise amount and rate of future warming that we can expect from rising greenhouse gas concentrations in the atmosphere is difficult to project, especially for specific locations. Projections of climate futures depend upon trends and estimates of greenhouse gas emissions but also on the sensitivity of the climate system to increased radiative forcing and the potential for positive or negative feedbacks. Positive feedbacks, such as decreases in the surface reflectivity or albedo, reinforce or amplify warming. Another example of a positive feedback is the melting of permafrost in Arctic tundra ecosystems, which increases emissions of methane, a potent greenhouse gas (Nauta et al. 2015). By contrast, negative feedbacks, such as increases in atmospheric aerosols, offset or reduce warming.

In order to understand what future climate change might look like, climate scientists have developed projections based on different future scenarios of greenhouse gas emissions, where higher concentrations imply greater radiative forcing and hence warming. While the earliest **global climate models** focused primarily on atmosphere–ocean processes, these models have become more complex over the years, and many of them now include representations of the land surface, ice sheets, clouds, aerosols, and other factors. To create future climate projections, scientists typically change some of the parameters within the models and observe the results over time in

Box 2.3 Radiative forcing and global warming potential

Discussions of the greenhouse effect often refer to the connections between the Earth's energy balance, **radiative forcing**, and global warming potential. These concepts help to explain how particular gases contribute to a warming planet.

Measurements of radiative forcing allow scientists to estimate how much different greenhouse gases have contributed to changes in the Earth's radiation balance. As an example, the sun's radiative forcing on the Earth's surface is 240 Wm^{-2} (watts per square meter). The addition of CO_2 into the atmosphere between the years 1750 and 2000 amounts to about 1.5 Wm^{-2}, and other gases (chlorofluorocarbons and hydrofluorocarbons, methane, nitrous oxide, ozone) total another 1.5 Wm^{-2}. In the short term, the positive forcing of CO_2 and other greenhouse gases is partly offset by aerosols (e.g., soot, pollution) that are released into the atmosphere from the burning of fossil fuels. These aerosols reflect some sunlight and have a cooling effect, which offsets the effects of some of the greenhouse gas warming. However, aerosols only remain in the atmosphere for a short time (a few days to a few years). A volcanic eruption like Mt Pinatubo (which occurred in the Philippines in 1991) can cause a negative forcing of 4 Wm^{-2}, but this effect lasts for only about two years. The net human climate forcing is currently estimated to be about 2.3 Wm^{-2} (IPCC 2013b). While this value may sound small in comparison to the radiative forcing of the sun, it is enough to contribute to the warming of the planet.

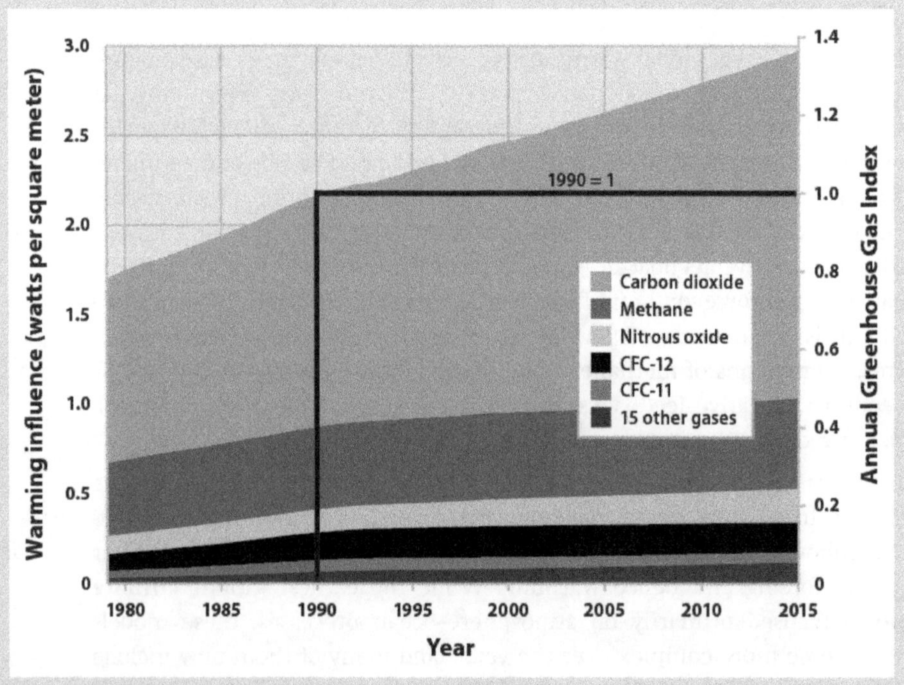

Figure 2.8 Radiative forcing of CO_2 and other gases.
Source: NOAA 2016

Global warming potential (GWP) is a simple index to compare how different gases contribute to global warming based on how much energy each gas absorbs over a set time period (typically 100 years). It measures how much heat a certain amount of gas traps in the atmosphere in comparison to how much heat is trapped by the same amount of CO_2. GWP depends on how well a gas absorbs energy and how long it remains in the atmosphere. Calculations of GWP are useful because they help to pinpoint which gases are the most potent contributors to global warming. Estimates of the net contributions to global warming depend not only on the GWP of a particular gas but also on the total quantity of that particular gas in the atmosphere. Methane gas (CH_4), for example, has a GWP of roughly 28–36 times that of CO_2 (USEPA 2016). Sources of methane emissions include ruminant animals (e.g., cows and other animals that digest grasses), rice paddies, and production and transport of fossil fuels. While emissions of CO_2 far exceed methane, methane is a particular concern because large quantities are stored within the soil and permafrost, and there is evidence that these stores are being released as soils warm and permafrost melt (Dean et al. 2018). As the oceans warm, there is also growing risk of large methane releases from ocean-floor storage. Although methane stays in the atmosphere for only a decade or two, its contribution to global warming is significant (Saunois et al. 2016).

order to explore how sensitive the climate system is to changes in greenhouse gas concentrations. Climate scientists are thus able to explore the role of greenhouse gases relative to other forcings, such as volcanic activities or changes in solar radiation, as well as the role of various feedbacks in dampening or amplifying the effects of climate change. The feedback effects have been particularly difficult to model, as some types of clouds have a cooling effect and others have a warming effect (Bony et al. 2015). However, a study by Steffen et al. (2018) suggests that biophysical feedbacks within the Earth system could play a more important role than earlier assumed, which would limit the range of potential future trajectories, introducing the possibility of a "hothouse Earth" scenario.

While global climate models have become considerably more complex over the past few decades, the projections of future temperature increases have remained relatively consistent. The first IPCC assessment report, published in 1990, projected temperature increases of 3°C by 2100 under "business as usual" **emissions scenarios** (IPCC 1990). This can be compared to projections from the Fifth Assessment Report in 2014, which suggests warming from 2.6°C–4.8°C by 2100 under a scenario of continued high emissions (see Figure 2.9).

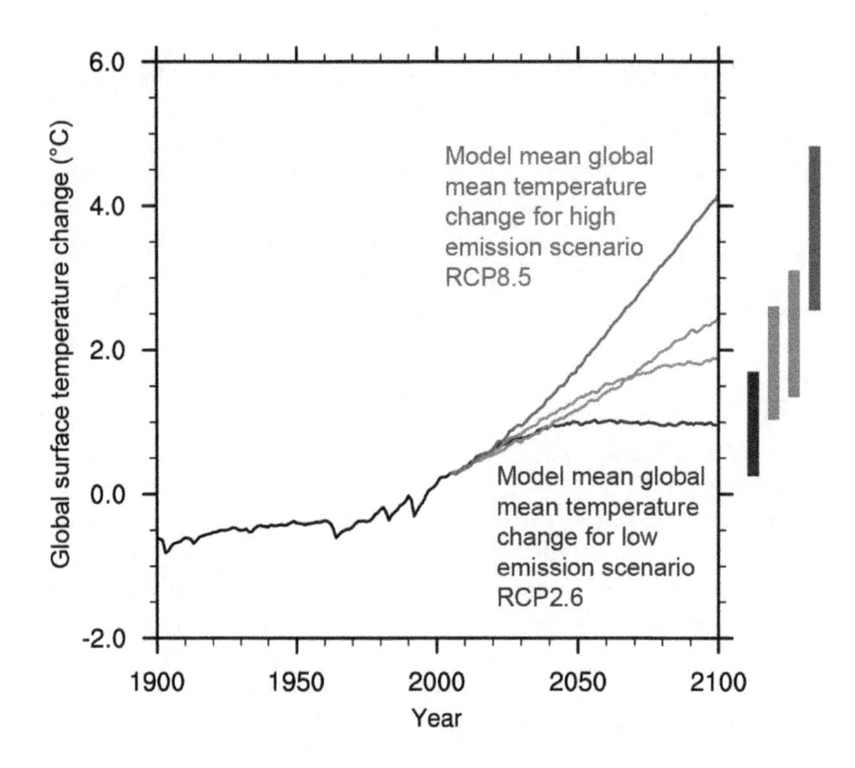

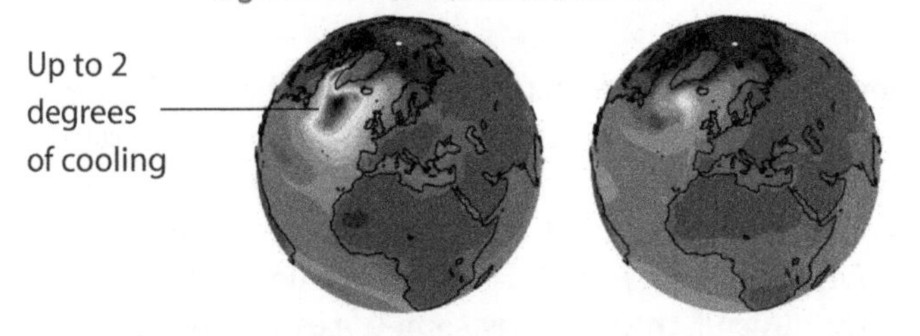

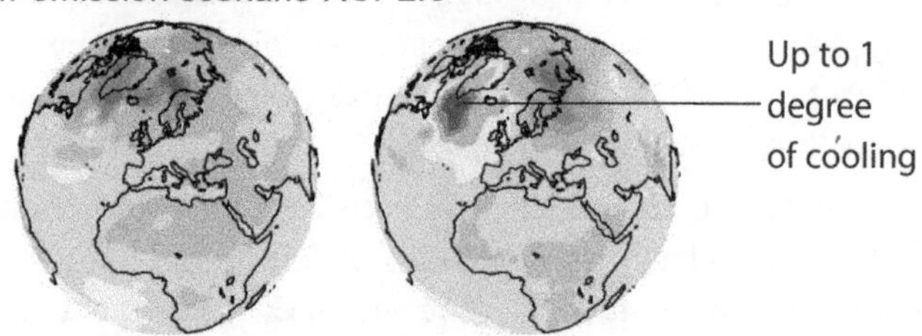

The darker the shading, the greater the increase in temperature, except for the marked regions

Figure 2.9 IPCC projections under different emissions scenarios.
Source: Based on IPCC 2013b

In thinking about climate futures, it is important to recognize that we are discussing climate projections here, as distinct from predictions. Climate projections are dependent on the assumptions in the models and also on scenarios of future greenhouse gas emissions. Projections of future climates include assumptions about the development and spread of renewable energy, climate and energy policies, and how society will be organized in the future. They also depend on assumptions about population growth, economic growth, and the types and distribution of economic activities. The more realistic projections will capture how much and how fast greenhouse gas emissions rise or fall over the next century.

Limiting global warming to below 2°C, which is consistent with the commitments made by national governments in the Paris Agreement, requires drastic reductions in greenhouse gas emissions. While it is easy to look at the different scenarios and conclude that a few degrees of change do not make a difference, the implications of each degree of change are not identical. For example, the consequences of an additional degree of temperature change can trigger positive feedbacks, such as the melting of sea ice (which changes the reflectivity of the surface and adds to warming). The uptake of carbon by oceans and forest may decrease in a warmer climate, creating another positive feedback. As mentioned above, cloud cover can have both amplifying and dampening effects, either warming the Earth's surface more or cooling it off. Despite uncertainties related to future climate projections, especially in relation to extreme events, the consequences of a global average warming of 1.5°C versus 4°C or more in the twenty-first century have profound implications for climate change impacts and adaptation, as we discuss in later chapters.

Returning to last chapter's discussion of geological time, atmospheric carbon dioxide levels at the beginning of the twenty-first century have reached levels that have not been experienced on the planet for more than 800,000 years (WMO 2017). Scientific evidence indicates that temperatures were much higher during geological epochs such as the Miocene (10–14 million years ago) when CO_2 concentrations exceeded 400 ppm. Evidence also suggests that, during such periods, the world was ice-free, and sea levels were many meters higher than they are today (Tripati, Roberts, and Eagle 2009). Study of past sea levels offers important insights into how much sea levels may be expected to change as temperatures rise in the future, and how rapidly such changes could take place. Based on studies of past climates and sea-level records, scientists estimate that over the next 2,000 years, sea levels will rise approximately 2.3 meters for every 1°C of temperature increase (Levermann et al. 2013). Within this century, sea levels are projected to rise by between 40 cm and 2 meters, depending on the rate, amount, and timing of melting from the Greenland and Antarctic ice sheets (IPCC 2014a).

In considering future projections, there are several important points to emphasize. First, it is the cumulative emissions that determine the temperature response in the atmosphere, rather than annual emissions. For this reason, climate scientists have paid considerable attention to the **carbon budget**, an accounting of how much carbon dioxide can be added to the atmosphere to meet different temperature targets (see Box 2.4). Second, the rate of change will also have consequences for the timing of impacts and adaptation. Third, projected global average temperature changes do not reflect the spatial distribution of climate change. Land areas will warm more than oceans, and higher latitudes will warm more than lower latitudes. There will also be temporal differences, with winter temperatures in the Arctic warming three times faster than average temperatures (New et al. 2011). Finally, the magnitude and timing of changes in the climate system will intersect and interact with social changes, which may exacerbate or reduce vulnerability and risk.

Box 2.4 Global carbon budget

One way to think about the connections between CO_2 emissions and global temperatures is through the concept of a global carbon budget. The carbon budget is based on the near-linear relationship between total carbon dioxide emissions and global mean temperature change: the more CO_2 that is emitted, the more that temperature increases. To keep future global mean temperature increases below a particular threshold, such as 1.5°C or 2°C, it is cumulative CO_2 emissions that are most relevant. Based on the carbon budget, limits for total emissions can be identified and serve as targets for policies and actions to reduce climate change impacts and risks. This means that past, current, and future emissions from the main sources of CO_2 (e.g., fossil fuel energy use, cement production, and land-use changes, including deforestation) have to be accounted for. Every ton of CO_2 emitted influences the remaining quota.

To limit global average temperature increases to less than 2°C, scientists calculate that a total carbon budget of no more than 2,900 Gt (gigatons) of carbon can be emitted into the atmosphere. It is estimated that more than 65% of this carbon budget has been used already. The remaining 35% (roughly 600–800 Gt carbon) represents only a fraction of known reserves of fossil fuels (IPCC 2014a). One study shows that "globally, a third of oil reserves, half of gas reserves and over 80 per cent of current coal reserves should remain unused from 2010 to 2050 in order to meet the target of 2°C" (McGlade and Ekins 2015: 187). Uncertainties associated with potential feedbacks also have to be taken into account.

The question of how to remain within a carbon budget of 600–800 Gt is central to many efforts to address climate change. Most scenarios associated with limiting warming to below 2°C assume a large-scale and successful use of **carbon sequestration** technologies, such as carbon capture and storage. They also assume significant reductions in deforestation and in emissions from cement production and may ultimately require "virtual elimination of CO_2 from the energy system" (Anderson 2015: 899).

Do you think that the concept of a carbon budget is useful for explaining the connections between rising temperatures, greenhouse emissions, and human activity? Do you think that keeping within a carbon budget that limits warming to less than 2°C is feasible?

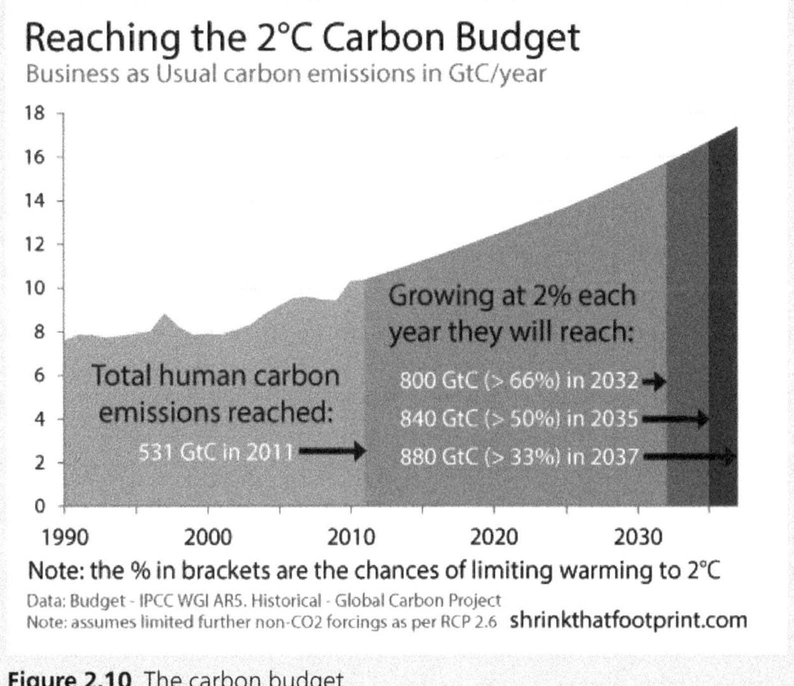

Figure 2.10 The carbon budget.
Source: http://shrinkthatfootprint.com/

Is science enough?

A basic understanding of climate science is critical to informed discussions about drivers, impacts, and responses to climate change. Science provides clear evidence that human activities are influencing the global climate system. Thousands of scientific papers have produced detailed evidence of changing temperature and precipitation patterns, sea-level rise, ocean acidification, changes in extreme events, and other consequences of increasing concentrations of greenhouse gases in the atmosphere. New research continually documents human impacts on marine ecosystems, on agriculture and forestry, in the Arctic, and in urban areas. Given the overwhelming scientific evidence that humans are influencing the Earth's climate, and that these ongoing changes will present a grave danger in the future, one might ask, "What are we waiting for?"; "Why aren't we taking action?"; and "Isn't the science enough?"

In contrast to the scientific consensus on climate change, public understanding of the issue remains much more ambiguous (Boykoff 2011). Many people still question whether climate change is real, whether it is caused by humans, or whether it is actually a problem. Within the United States, for example, recent research indicates that one in six people (16%) think global warming is not happening, whereas one in three believe it is due to natural variability (Leiserowitz et al. 2015). These figures are often contrasted with the scientific consensus on climate change, which indicates that, among climate scientists, more than 97% think that recent global warming is caused by humans (Cook et al. 2016).

While there are many strategies for enhancing public understanding and engagement with climate change, one area that has drawn considerable attention is the communication of the science itself. The use of scientific findings and information represents a particular mode of communication about climate change. This mode makes use of text, tables, figures, graphics, and animations to illustrate key findings of scientific reports and papers. It is premised on an **information deficit model** of communication, which presumes that audiences – whether students or members of the public – need to better understand the science in order to be convinced that climate change is a real and pressing concern. The assumption is that better knowledge of the science will lead to more engagement and action (Moser and Dilling 2011).

Although the idea of a climate change information deficit may seem logical, studies in psychology show that many people are not convinced about something through scientific data alone (Weber 2010). As we discuss in later chapters, people are often more responsive to stories and emotions, rather than facts and cognition. Seeing a picture of a starving polar bear or hearing about children who have lost their homes in a storm may be a more powerful way of communicating climate change than through reports with colorful graphs or maps (O'Neill 2018). Moreover, people are social beings and tend to share the attitudes and beliefs of their friends and those within their social circles (Kahan 2012). Although more research is vital for understanding human impacts on the climate system, it is unlikely that such information alone will be effective in mobilizing society to respond to climate change or to create the social and political momentum to support more sustainable approaches to development.

If better knowledge and information about climate change are not sufficient to raise concerns that result in action, what are the alternatives? Climate communication is an evolving field of scientific study, and recent work suggests a number of alternatives to the information deficit model. In particular, dialogue and participatory processes have proven to be effective approaches for engaging people with climate change (Gramberger et al. 2015). Such approaches place greater emphasis on discussing climate change, rather than informing or convincing people about it. Recent work on this theme also points to the importance of understanding different audiences and framing the information in ways that are consistent with the audience's beliefs, values, and worldviews (Moser 2016). Researchers emphasize the key role of narratives, imagery, and visualizations in climate communication, as well as the need to connect with people's emotions and identities (Veland et al. 2018). Yet, as researchers Moser and Dilling (2011) point out, in order for communication to be effective in contributing to active engagement, it must also be supported by policies, incentives, and infrastructures that allow audience concerns to be addressed.

Summary

Pulling together the information presented in this chapter, we can summarize the scientific narrative on climate change in a rather simple way: The climate is changing and it is recognized to be influenced by human activities. Significant changes are projected for the future and, as we discuss in later chapters, they will have profound impacts on society. Human decisions, innovations, and actions can play a key role in minimizing climate change. While scientific evidence provides strong justification for the need to take action to reduce emissions of greenhouse gases, scientific knowledge alone is not sufficient to generate social responses, at least not on the scale and at the rate that is needed to avoid dangerous climate change. In the next chapter, we discuss the importance of discourses in framing both the problem and the solutions.

Reflection questions

1 How would you explain the evidence of climate change to a friend, relative, or "person on the street" who may be confused about the difference between climate and weather?
2 Why isn't the scientific consensus on climate change "enough" to motivate responses?
3 Besides more scientific evidence and information, what else might inspire action on climate change?

Further reading

Boykoff, M. T., 2011. *Who Speaks for the Climate? Making Sense of Media Reporting on Climate Change*. Cambridge, UK: Cambridge University Press.

Cook, J., Oreskes, N., Doran, P. T., et al., 2016. Consensus on consensus: A synthesis of consensus estimates on human-caused global warming. *Environmental Research Letters* 11(4): 048002.

Hansen, J., Sato, M., and Ruedy, R., 2012. Perception of climate change. *Proceedings of the National Academy of Sciences* 109(37): E2415–23.

IPCC, 2014a. *Climate Change 2014: Synthesis Report. Contribution of Working Groups I, II and III to the Fifth Assessment Report of the Intergovernmental Panel on Climate Change*. Geneva: IPCC.

Moser, S. C. and Dilling, L., 2011. Communicating climate change: Closing the science–action gap. *The Oxford Handbook of Climate Change and Society*. Oxford: Oxford University Press, 161–74.

3 CLIMATE CHANGE DISCOURSES

What kind of a problem is climate change?

Climate change is often described as the greatest problem facing humanity. But what type of problem is climate change, actually? This is an important question, since how we define a problem influences the types of solutions that will be identified and prioritized. Most commonly, climate change is understood as an environmental problem related to increasing levels of greenhouse gases in the atmosphere as a result of human activities. Although knowledge of the science of climate change is critical for understanding the physical basis of the problem, viewing climate change as an environmental problem leads to prioritizing particular types of solutions, particularly technical measures and policy interventions to limit greenhouse gas emissions and adapt to the impacts.

While climate change may indeed be considered an environmental problem, there are many other ways to approach it. In this chapter, we show that climate change is not a monolithic issue, but one that can be seen through different lenses. For example, some interpretations suggest that the problem is the result of economic, political, and cultural forces that are perpetuating fossil fuel development and consumption, leading to rising emissions. Others dismiss the problem of climate change, labeling it as a scam or hoax that is intended to promote government regulation or increased funding for science. Still others see the problem of climate change as deeply rooted in particular beliefs and perceptions of human–environment relationships and humanity's place in the world, which influences norms, rules, and institutions that support unsustainable resource use and practices.

We begin the chapter by describing how discourses shape the ways in which climate change is defined and studied, as well as the types of responses that are identified and prioritized. We then describe four types of discourse on climate change, showing how each represents a particular approach to understanding the issue and consequently supports certain types of strategies for addressing it. We will draw on this discussion of discourses in subsequent chapters as we present different understandings of the drivers, impacts, and responses to climate change.

The power of discourses

> . . . we have to live with the fact that different individuals and groups use different discourses to make sense of the same nature/s. These discourses do not reveal or hide the truths of nature but, rather, *create their own truths*. Whose discourse is accepted as being truthful is a question of social struggle and power politics.
>
> Castree 2001: 12

A discourse may be understood as a system of representation that is made up of norms, rules of conduct, institutions, and language that influence and legitimize certain perspectives and meanings over others (Leichenko and O'Brien 2008). Discourses include explicit and implicit values, judgments, and contentions that define the terms of discussion around a particular issue, as well as what is included and excluded from analysis and debate. Discourses reflect different perceptions and understandings of climate change, which are linked to world-views, assumptions, and beliefs about both nature and society (see chapter 4).

Narratives and stories are considered essential components of discourses (Veland et al. 2018). In interpreting the causes, consequences, and responses to climate change, each discourse reflects different understandings of causality and possibilities for societal change. Clues to discourses on climate change lie in the language and texts that are used to talk about the issue and its potential solutions. Words such as "urgent" and "irreversible" may dominate some discourses, whereas others might focus on words such as "uncertainty" and "doubt." Still others highlight concepts such as "equity" and "justice" or emphasize notions of "connection" and "integration." The language and objects of interest within a particular discourse are likely to appeal to those who are aligned with it while being ignored or rejected by those who are part of other discourses.

Some discourses carry considerable political weight and may support existing interests and power structures, while others specifically challenge them (Castree 2001). In practical terms, discourses influence the way that climate change is addressed. Those who approach climate change as an environmental issue are likely to emphasize mitigation policies as solutions. Those who see it as an issue of development may call attention to the unfair industrial, agricultural, and trade policies and practices that contribute to risks and vulnerabilities. People and organizations that view climate change as a conspiracy are likely to distrust efforts to address it through international agreements, national policies, and local actions. Businesses that view climate change as an opportunity may recognize the potential to develop new markets for sustainable and energy-efficient products and services.

By drawing attention to discourses, we see that different approaches to climate change are rooted in particular ways of making sense of the world.

However, discourses are also flexible, porous, and sometimes overlapping, and they can change in response to social, economic, cultural, and political circumstances, as well as to shifts in values and priorities, especially among younger generations. In the following sections, we describe four broad discourses on climate change and society: biophysical, critical, dismissive, and integrative (see Table 3.1). Both the labels and content of the discourses that we describe are intended to serve as wide umbrellas that capture and illustrate some of the distinguishing features of the many different ways that people make sense of climate change. These discourses, which build upon and extend the description of global change discourses presented by Leichenko and O'Brien (2008), are not intended to be comprehensive, fixed, or all-inclusive. Indeed, many other environmental discourses and classifications can be identified (see Dryzek 2013). By presenting a general typology of climate change discourses, our intention is to illustrate their power in shaping understandings, discussions, and debates about the issue.

Table 3.1 Four types of climate change discourses

Biophysical discourse Climate change is an environmental problem caused by rising concentrations of greenhouse gases from human activities. Climate change can be addressed through policies, technologies, and behavioral changes that reduce greenhouse gas emissions and support adaptation.
Critical discourse Climate change is a social problem caused by economic, political, and cultural processes that contribute to uneven and unsustainable patterns of development and energy usage. Addressing climate change requires challenging economic systems and power structures that perpetuate high levels of fossil fuel consumption.
Dismissive discourse Climate change is not a problem at all or at least not an urgent concern. No action is needed to address climate change, and other issues should be prioritized.
Integrative discourse Climate change is an environmental and social problem that is rooted in particular beliefs and perceptions of human–environment relationships and humanity's place in the world. Addressing climate change requires challenging mindsets, norms, rules, institutions, and policies that support unsustainable resource use and practices.

The biophysical discourse

> Continued emissions of greenhouse gases will cause further warming and changes in all components of the climate system. Limiting climate change will require substantial and sustained reductions of greenhouse gas emissions.
>
> IPCC 2013b: 19

The most prevalent discourse on climate change is what we term the "biophysical discourse." Within this discourse, climate change is understood in environmental terms, as a problem linked to rising concentrations of greenhouse gas emissions from human activities. A common narrative points to the ways that humans have collectively contributed to climate change. This narrative tends to assign shared responsibility for climate change to all humans. Those working within the biophysical discourse draw heavily on Earth-system science research to understand how human activities such as fossil fuel consumption, deforestation, and other land-use changes influence the atmosphere, biosphere, oceans, and ice cover (Steffen et al. 2005) (see Box 3.1). Understanding relationships among these processes is key to identifying the consequences of climate change for natural and social systems.

Within the biophysical discourse, an important priority is to explore "what if" scenarios of climate change. For example, what will happen to global temperatures if carbon dioxide emissions continue to increase at the current rate over the remainder of this century? How much will global sea levels rise if melting of the Greenland ice sheet accelerates? How will these changes influence particular places or populations? Exploration of these issues often makes use of scientific tools such as global climate models to investigate the short- and long-term impacts of increasing greenhouse gas concentrations on the climate system (McGuffie and Henderson-Sellers 2005). These models are also used to project how the climate will change within different regions under different scenarios of greenhouse gas emissions. These projections are typically translated into impacts on human and natural systems, as we will discuss in later chapters.

Although we did not emphasize the notion of a biophysical discourse in chapter 1, the description of the Anthropocene as a new geological epoch in which human activities are pushing planetary boundaries and threatening to disrupt Earth-system processes is characteristic of this discourse. The biophysical discourse was also dominant in chapter 2's discussion of the scientific basis of climate change. Common to the biophysical discourse is an approach to scientific knowledge that prioritizes **positivist science**, whereby physical processes and interactions are objectively observed, understood, modeled, and validated. The biophysical discourse is premised on the assumption that greater scientific knowledge will reduce uncertainty about the consequences of increased greenhouse gas emissions, and that a basic social consensus on climate policy responses and actions will follow. This reflects the information deficit model of communication, which assumes that more knowledge and better science communication will translate into policy and action to solve environmental challenges (Moser 2016). Not surprisingly, there are often calls within the biophysical discourse for more funding to support climate change research, including development of more detailed models and scenarios to investigate potential climate futures.

Box 3.1 Earth-system science and the Bretherton diagram

The biophysical discourse on climate change is largely rooted in the field of Earth-system science. This field developed during the 1970s and 1980s as an outcome of advances in systems modeling and systems thinking. The approach draws attention to relationships between different components or subsystems of a larger biogeophysical system. An early view of how human activities were connected to larger systems is encapsulated in the well-known Bretherton diagram of the Earth system (see Figure 3.1).

In this diagram (originally developed by Professor Frances P. Bretherton), we see that human activities are depicted as a separate box that influences and is impacted by biophysical processes. This "wiring diagram" emphasizes that the physical climate system is linked to biogeochemical cycles (nitrogen, phosphorus, carbon, and sulfur) and to terrestrial and ocean systems. As Brulle and Dunlap (2015) note, from a natural science perspective, all human activities are collapsed into a black box. This may explain why the social sciences, and the critical discourse in particular, have been marginalized in climate change research and policy.

What does the placement of human activities into a separate "oval" suggest about the perceived role of humans within the Earth system? What role do you see for human agency within this diagram?

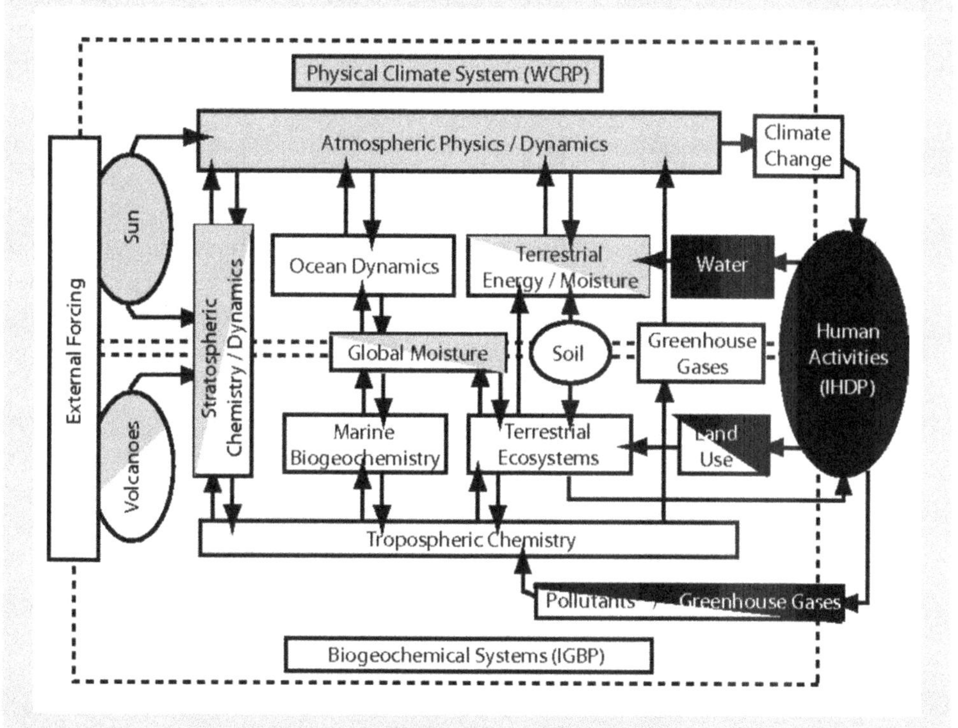

Figure 3.1 A simplified version of the Bretherton diagram.
Source: Mooney, Duraiappah, and Larigauderie 2013

The science–policy interface receives considerable attention in the biophysical discourse. The connections between scientific knowledge and formation of climate policy are exemplified in the assessments carried out by the Intergovernmental Panel on Climate Change (IPCC). The IPCC is a scientific and intergovernmental body that was established in 1988 to "provide the world with a clear scientific view on the current state of knowledge in climate change and its potential environmental and socio-economic impacts" (IPCC 2018a). Governments participate in the review process, as well as in plenary sessions where decisions about the programs and priorities are decided, and where the reports are presented, discussed, revised, and approved. The general guidelines for IPCC reports emphasize that results should be "policy relevant, but not policy prescriptive" (Rothman et al. 2009). However, debates during IPCC plenary sessions, which include both scientists and government representatives, sometimes reveal tensions between the interpretation of scientific information and national political interests. Indeed, much is at stake in these deliberations, as IPCC findings are a key scientific input for international climate frameworks and agreements, including those negotiated under the UN Framework Convention on Climate Change (UNFCCC). The scientific basis for the 2°C and 1.5°C targets under the Paris Agreement are frequently discussed in relation to the biophysical and political feasibility of reaching them (Friedlingstein et al. 2014). However, as Knutti et al. (2016: 1) point out "the 2°C target is a political consensus that takes into account what policymakers at that time considered to be both realistically achievable and tolerable."

The biophysical discourse uses tools such as climate models and economic models to support what is sometimes referred to as a **techno-managerial** approach to climate change responses. Indeed, within the biophysical discourse, technological innovations, such as development and deployment of non-fossil fuel energy sources, increasing the energy efficiency of vehicles or consumer products, or development of drought-resistant seeds, are often seen as key strategies for climate change mitigation and adaptation (Meinshausen et al. 2009; Biagini et al. 2014). Managerial solutions might include regulations that limit emissions from power plants, guidelines for new constructions in flood-prone areas, or policies that ration water in times of drought. Policy innovations might also entail developing greenhouse gas inventories for countries or firms and new methods for counting, tracking, and monitoring emissions (Torres, Andrade, and Gomes 2017).

Within the biophysical discourse, mitigation actions also tend to emphasize the role of individual behavioral changes. More sustainable practices, such as walking, riding a bike, or taking public transportation to work instead of driving a car, are seen as fundamental to reducing greenhouse gas emissions. These practices can be motivated through **market-based measures** and incentives or through **nudging** behaviors. Such strategies may include, for example, putting a tax on carbon so that it becomes more expensive to purchase fossil fuel energy or providing free public transit passes to employees (Lieberoth, Holm Jensen, and Bredahl 2018). Though there is a strong emphasis on individual action, it is often through appeals to what Elizabeth Shove (2010) refers to as "attitude, behavior, and choice" rather than through collective action, social struggle, and political change. As we discuss in the next section, an emphasis on social processes leads to different framings of the climate change problem, as well as different types of responses.

The critical discourse

> [T]he three policy pillars of the neoliberal age – privatization of the public sphere, deregulation of the corporate sector, and the lowering of income and corporate taxes, paid for with cuts to public spending – are each incompatible with many of the actions we must take to bring our emissions to safe levels.
>
> Klein 2015: 72

The critical discourse draws attention to the role of political, economic, and cultural processes and social relations in shaping drivers, impacts, and responses to climate change. Within this discourse, narratives emphasize the highly uneven contributions to climate change across society, drawing attention to politics, power, and vested interests in shaping planetary futures. We refer to this as the critical discourse for two reasons: (1) it questions the dominant (i.e., biophysical) discourse on climate change; and (2) it draws upon social theory to present a broader critique of society, with the goal of changing dominant systems and structures. Over the past decade, critical perspectives on climate change have become more visible and prevalent, especially within the social science research community.

The critical discourse has been influential in highlighting how social and economic processes and power relations are driving climate change. In particular, it draws attention to the role of the capitalist economic system, sometimes referred to as "fossil fuel capitalism," in fostering and perpetuating an economic growth imperative, which in turn leads to unsustainable patterns of resource use (Wilhite 2016). Social activist and author Naomi Klein (2015), quoted above, argues that climate change is actually a crisis of capitalism and that it can be used as wake-up call to address problems in the dominant economic system. Those within the critical discourse question the biophysical framing of the Anthropocene and suggest that a label such as "Capitalocene" might be more appropriate for describing the new epoch (Haraway 2015; Altvater et al. 2016).

Common to approaches within the critical discourse is the recognition that all humans are not equally responsible for climate change, nor will they all experience the consequences in the same way (Liverman 2009). Equity and social justice issues are seen to be at the core of climate change, and this results in calls for social and political responses, as well as technical ones (Klinsky et al. 2017). As such, the critical discourse raises questions about whose voices are heard, who makes the rules, whose values count, and who decides the future. The discourse emphasizes the importance of wide participation in science-policy process to ensure that a diversity of perspectives is included in climate change-related decisions. As an alternative to the notion of planetary boundaries and a safe operating space for humanity, some researchers within the critical discourse propose the idea of "safe and *just* operating space for humanity" (Dearing et al. 2014).

The critical discourse has been influenced by **environmental justice** movements, which have drawn attention to inequalities in exposure to environmental hazards based on race, gender, ethnicity, and income (Cole and Foster 2001; Cutter, Ash, and Emrich 2014). These movements have emphasized the need for more transparent, inclusive, and equitable decision making around environmental issues. The critical discourse recognizes how historical legacies of colonialism, unequal political and economic power, and unfair economic and trade policies, combined with inequalities on the basis of race, class, gender, indigeneity, and other factors, play a decisive role in influencing both vulnerability and the capacity to adapt to climate change (Cameron 2012). Critical approaches also illustrate the diverse ways that socioeconomic conditions and policies shape climate change risks. These approaches also explore the ways in which certain behaviors, such as driving an automobile or eating meat, are internalized and habituated through social norms, rules, and incentives (Hargreaves 2011). Critical approaches recognize that without attention to social, economic, and political contexts, including politics and power relationships, an emphasis on technical and behavioral responses is unlikely to contribute to the types of structural and systemic changes needed to address the underlying causes of climate change. Indeed, solutions to climate change from the perspective of the critical discourse stress the need to challenge economic, political, and cultural processes that contribute to climate change.

In contrast to the biophysical discourse, which tends to emphasize technological, behavioral, and policy changes as solutions, the critical discourse points to the need to address **root causes** of risk and vulnerability (Ribot 2014). This implies challenging economic systems that are seen as unfairly advantageous to some over others, addressing unequal access to and distribution of resources, confronting the role of politics, power, and the vested interests in particular forms of development, and, above all, questioning the status quo. The discourse is more explicitly political than the biophysical discourse, in that it emphasizes individual and collective capacities to challenge and transform current economic and social development trajectories.

Proponents of the critical discourse question positivist approaches to science, advocating instead what is sometimes described as **post-normal science** (Ravetz 2006). Arguing that science is neither neutral nor unbiased, the critical discourse reminds us that scientific questions and methods are always influenced by historical and contemporary political, economic, and cultural contexts, as well as by the perspectives and biases of the researchers themselves. Such critiques do not question the physical reality of climate change but instead point out that scientific methods and approaches represent only one way of seeing the world (Hulme 2011). An emphasis on the social construction of knowledge recognizes that there are different ways of both knowing and defining what is valid or true, which in turn helps to explain the dismissal or denial of climate change as a serious problem.

Dismissive discourse

> It is the greatest scam in history. I am amazed, appalled and highly offended by it. Global warming . . . it is a SCAM.
>
> John Coleman, Founder of the Weather Channel, 2007: 2

Up to this point, our discussion has focused on two discourses that accept climate change as a real and pressing issue. A third type of discourse questions the reality and significance of climate change. The dismissive discourse encompasses a diversity of views that trivialize or deny climate change as a problem linked to humans and their activities. In particular, it downplays the significance of climate change science and the need for rapid and widespread responses to address risk and vulnerability and to promote sustainability. In some cases, the dismissal is linked to beliefs, including the belief that humans cannot possibly change the climate because such systems are beyond human influence, or because only God or a deity can intervene in the climate. In other cases, it is linked to economic interests of individuals or industries (see Box 3.2). The dismissive discourse influences political and cultural debates about climate change policy and action, and it affects how climate change is portrayed in the media and in many educational settings.

While this discourse is often reduced to a matter of believing or not believing in climate change, there are, in fact, many ways of dismissing the issue. Below we identify and discuss three broad categories that constitute the dismissive discourse on climate change. In identifying these categories, we draw from the work of Kari Norgaard (2011) on the social organization of climate change denial. Our categories include those who see climate change as a hoax, conspiracy, or scandal, those who are aware of climate change but do not believe that it is caused by humans, and those who accept the science of climate change but do not consider it to be a serious problem. Common

Box 3.2 Skepticism and the politicization of scientific information

While uncertainties about future climate trajectories are sometimes viewed by climate skeptics as "proof" that climate change is not real or not something that we need to worry about, skepticism is actually an important part of the scientific enterprise. Skepticism of research findings or conclusions is vital for identifying weaknesses in theories or methods. Global climate models, for example, have developed over time to include more processes and details, partly in response to questions raised by skeptics about the differences between the results of model simulations of the climate as compared to observations and historical data. This type of skepticism has improved the accuracy of the model projections.

Many researchers are concerned about what they consider to be the "politicization" of climate change science. Those within the dismissive discourse often make the case that climate change research is biased and sometimes accuse climate change scientists of making up their results because they have a hidden agenda, such as promoting progressive policies or increasing funding for science.

Although there is little evidence to support the notion of a large-scale conspiracy among climate change scientists, there are many examples in which the deliberate obfuscation of scientific information has been used to serve political interests. The 2014 documentary film *Merchants of Doubt*, based on a book of the same name by Naomi Oreskes and Erik M. Conway (Oreskes and Conway 2010), examines how scientific evidence about the safety of products such as tobacco and fire retardants has been intentionally distorted or hidden in order to serve the economic interests of particular industries. The documentary shows that similar strategies have been used by industry groups to raise public doubts about the reality of climate change. This includes paying prominent "contrarian" scientists to question the scientific consensus, or hiding information about known dangers associated with global warming. While the film's message is closely aligned with the critical discourse, it illustrates how science can be interpreted (or deliberately misinterpreted) for political purposes: a point that is relevant to understanding all climate change discourses.

Can you think of another issue where scientific evidence has been used for political aims or the science has become "politicized"? Does this issue offer any lessons or insights for understanding debates and disagreements about climate change?

to each of these categories is the tendency to downplay the significance of climate change and its implications for the future.

1 *Dismissing the science of climate change*. The scientific evidence of climate change is looked upon with suspicion by people who do not accept science as a valid mode of inquiry or who believe that scientists are lying (see Box 3.2). They often consider scientific claims as a means to realize hidden objectives. This may include efforts to promote interventions to increase the size and scope of government, whether by monitoring and regulating carbon emissions, by imposing greater government control over energy production, or by prioritizing policies that favor the natural environment and non-human species over humans. A major concern among **climate skeptic**s is that climate change will be used to promote a government "takeover" of the economy, or government involvement in behavioral decisions (e.g., regulation of the types of light bulbs that can be bought or sold). Some are convinced that scientists are using climate change doomsday scenarios to scare the public, which can then be used as justification to secure more research funding.

2 *Dismissing human causes of climate change*. Some people recognize that the climate is changing, but they dismiss the possibility that humans are responsible for the changes. As we discuss in chapter 4, the recognition that humans have the capacity to directly alter the climate system challenges some belief systems, especially those accepting that only their god or a higher force can influence the climate. Some people are convinced that humans are too small and insignificant to make a difference to the climate system, and instead they point to the role of sunspots, volcanic activity, and other "external" drivers as the cause of observed change. Those who argue that climate change is a natural process that is outside human control also doubt that humans have an ability to respond, either to accelerate or slow climate change. This group, described by some as "neo-skeptics," believes that humans cannot and should not do anything about climate change (Stern et al. 2016). Those holding this view may assume that development and progress will resolve any problems created by climate change. They may also assume that climate-related catastrophes are natural and inevitable and that humans cannot do anything to stop them. Like climate skeptics, some may feel that taking steps to address climate change through policy merely serves the interests of renewable energy firms, environmental activists, and others who support regulatory or policy efforts to limit greenhouse gas emissions.

3 *Dismissing the significance of climate change relative to other issues*. Some people accept the science of climate change and believe that it is occurring and is indeed a problem. However, they do not think the impacts of climate change will be universally significant, and they point to many other social and environmental problems as more pressing and urgent. Poverty, migration, substance abuse, human rights violations, and so on may be considered

higher priorities when ranked against climate change. Sometimes climate change is dismissed as a problem because the economic costs of taking action are considered to have negative effects on the economy. For example, economic arguments might be made that people in the future will be richer as a result of economic growth and rising economic productivity, thus better able to afford to mitigate climate change and to adapt to the impacts. Sometimes it is argued that climate change policies (e.g., shifting to alternative, non-fossil fuel energy sources) will slow efforts to reduce poverty in developing countries, or that the positive effects of climate change (e.g., lower heating costs, longer growing seasons) are more beneficial than the negative effects. Such positions may tacitly support the perception that wealthy populations and developed countries will be able to adapt to climate change impacts, or that technological solutions will eventually be found to sequester and reduce atmospheric concentrations of carbon dioxide.

Although we have highlighted some of the specific reasons for dismissing climate change, a more expansive interpretation of the dismissive discourse might also include people who recognize the significance of climate change, yet nonetheless dismiss the issue as an abstract, distant, or overwhelming problem that can be ignored in everyday life. As mentioned in chapter 1, Norgaard (2011) considers this type of socially organized dismissal as implicatory denial. This form of denial is not direct but implied, in that knowledge about climate change is accepted but is not translated into action. Many people are likely to experience implicatory denial in relation to climate change at some point in their lives, and it may be humbling to acknowledge that most humans today are "living in denial."

Although it may seem easy to pigeonhole those within the dismissive discourse as uninformed, misinformed, influenced by special interests, or simply hypocritical, it is important to recognize that this discourse holds considerable political influence. The power of discourses often lies in their ability to define the terms of the argument through a language and logic that appeals to and gains support from others. When a substantial segment of the population ignores or minimizes the importance of climate change, it can have significant political and practical implications. The decision taken in 2017 by the president of the United States to withdraw from the Paris Agreement, justified in part on the basis of skepticism about human causes of climate change, illustrates the power associated with the dismissive discourse.

The integrative discourse

[D]eliberately and collectively anticipating the future is a deeply historical and cultural process, drawing in our shifting ideas about time and

> certainty, humans and our relationship to the environment, and the desirability and inevitability of existing aspects of the world.
>
> Rickards et al. 2014: 598

The discourses described above represent three very different approaches to climate change. There is, however, also an emerging discourse on climate change that views the issue from a holistic perspective, which we refer to as the integrative discourse. Whereas the biophysical and critical discourses place primary emphasis on either physical or social systems, the integrative discourse sees climate change as interconnected with multiple processes of environmental, economic, political, and cultural change, and closely linked to individual and shared norms, beliefs, values, and worldviews (Gibson-Graham 2011). Bringing together insights from multiple perspectives, the integrative discourse approaches climate change as a transformative process. It is transforming the environment; it is transforming communities and cities; it is transforming how we perceive and relate to nature and each other; and it is transforming how we engage with the future. The potential to deliberately transform human–environment relationships in ways that are equitable and sustainable lies at the heart of the integrative discourse, as we will discuss in chapter 10.

The integrative discourse draws attention to the ways that subjective beliefs, values, and worldviews both reflect and influence social systems and human–environment relationships. In particular, the discourse questions **dualistic views** of nature as separate from society. This dualism reflects a particular pattern of western thought that can be largely traced to the Enlightenment – a period during the eighteenth century where scientific and philosophical ideas flourished, giving rise to the scientific method, as well as greater emphasis on rationality and individualism. Dualistic views are reflected in most of the social and material relationships associated with modernity, including the treatment of nature as an object to be managed, exploited, or preserved (Merchant 2005). This view may lead some scientists to advocate for more research on geoengineering the climate system through solar radiation management or carbon capture and storage – strategies that implicitly assume that the climate system is an external object that can be controlled and managed by humans.

There has been a long tradition of challenging the human–nature dualism in philosophy, geography, and feminist theory (Gerber 1997). Integrative approaches build on these traditions, which are also recognized by many within the critical discourse, to emphasize the unclear lines between what is considered "human" and "natural," and to redefine agency to include non-humans (Barad 2007; Dwiartama and Rosin 2014). The integrative discourse also recognizes that the questioning of paradigms and patterns of thought can foster new ways of seeing systems and solutions to complex, adaptive problems like climate change. The integrative discourse challenges linear and deterministic understandings of the connections between physical and social processes and impacts, recognizing that such understandings often support the conclusion that climate change is the most important driver of future social conditions and human destinies and that predicted climates will decide the future (Hulme 2011: 249). Viewing humans as a reflexive part of the climate system that is able to create, recognize, and change patterns introduces possibilities for changing the very relationships that create risk and vulnerability in the first place.

Integrative approaches to climate change emphasize that piecemeal solutions can exacerbate old problems or contribute to new ones (Olsson et al. 2017). For example, a focus on reducing carbon dioxide emissions may not address the political economy of land-use changes and resource practices that undermine livelihoods and threaten biodiversity. By the same token, reducing inequalities in resource access may not ensure that resources are managed in a sustainable manner. In terms of solutions, the integration of multiple perspectives can help to identify and generate new approaches to global challenges. This includes transdisciplinary approaches to mitigation, adaptation, and transformation (Ziervogel, Cowen, et al. 2016). It does not offer recipes or mandates for specific actions but instead opens the door for new inquiries and new ways of relating to both problems and solutions. The integrative discourse recognizes that social and cultural transformations are both possible and underway. Again, this is a theme that we will return to in the chapters ahead.

Summary

This chapter has explored the role of discourses in shaping how we understand the challenge of climate change. We have shown that discourses influence what people think about climate change, including perceptions of actions that can and should be taken in response. While all discourses are partial and prioritize some perspectives over others, the power of discourses should not be underestimated. Climate change discourses reflect different ways of seeing the world, different understandings of human–environment relationships, and different visions of the future. We explore these issues in the next chapter, showing how the discourses we relate to are deeply entangled with how we make sense of and relate to the world, both cognitively and emotionally.

Reflection questions

1 Which discourse best describes the way that you approach the issue of climate change? Why do you identify with this discourse?

2 Why does awareness of different discourses matter when it comes to communicating about climate change?
3 Does the recognition of different climate change discourses diminish the value of evidence-based scientific information about climate change? Why or why not?

Further reading

Castree, N., 2001. Socializing nature: Theory, practice, and politics, in N. Castree and B. Braun (eds), *Social Nature: Theory, Practice, and Politics*. Malden, MA: Blackwell, 1–21.

Dryzek, J. S., 2013. *The Politics of the Earth: Environmental Discourses*, 3rd edn. Oxford, UK: Oxford University Press.

Norgaard, K. M., 2011. *Living in Denial: Climate Change, Emotions, and Everyday Life*. Cambridge, MA: MIT Press.

Shove, E., 2010. Beyond the ABC: Climate change policy and theories of social change. *Environment and Planning A: Economy and Space* 42(6): 1273–85.

Ziervogel, G., Cowen, A., and Ziniades, J., 2016. Moving from adaptive to transformative capacity: Building foundations for inclusive, thriving, and regenerative urban settlements. *Sustainability* 8(9): 955.

4 WORLDVIEWS, BELIEFS, AND EMOTIONS

How do you see the world?

How do you see the world? Do you see the Earth and its environment as fragile and in need of protection, or do you see it as resilient, able to respond to and recover from damages related to human activities? Or maybe you don't see the environment as separate from humans at all? Do others share your views? What appears obvious to you might not be viewed the same by others. Your parents and grandparents, for example, might have different attitudes than you towards the environment and the issue of climate change. Future generations may have worldviews that are inconceivable to you and other people living today.

Worldviews often vary over generations, but they can also change through pivotal moments or experiences that trigger new ways of seeing and thinking about human–environment relationships. For example, one such moment occurred in 1968, when photographs of the Earth were taken by the crew of the Apollo 8 space mission (Figure 4.1). The spectacular image of "Earthrise" showed a green and blue planet with no visible political boundaries and borders. This unified view of the world, which first appeared in print in January 1969, changed the way that many people saw themselves in relation to the Earth and helped pave the way for the environmental movement and new laws, regulations, and international agreements to promote environmental sustainability.

Although that iconic image of planet Earth is now taken for granted, it is easy to forget that five hundred years ago, many people believed the Earth was flat. Two hundred years ago, the ability to travel, to meet people living in different environments, or to converse with people from different cultures and backgrounds was limited for most people. One hundred years ago, people sent letters that would take weeks to be delivered. Telegrams and telephone calls were new technologies that were available to relatively few people. Today, more than two-thirds of the global population has a cellular phone that can transmit and receive calls, messages, pictures, and information instantaneously to and from almost anywhere on Earth. Commercial aviation, which grew rapidly in the late twentieth century, now transports

Figure 4.1 Earthrise (Apollo 8), December 24, 1968.
Source: NASA

millions of people and goods around the world every day. Globalization of travel and technology, including expanded connectivity via the internet and social media, makes it easier to meet and interact with people from different cultural backgrounds and social contexts. Greater connectivity also means that we are frequently confronted with different worldviews.

In this chapter, we explore worldviews and consider how they relate to our understandings of climate change. We begin by defining worldviews and presenting some broad ways to categorize them. We then look at how worldviews shape understandings of nature and human–environment relationships. Next, we discuss beliefs, values, social norms, and emotions, including the ways that they influence how we perceive and relate to climate change. In the last section, we explore how artistic imaginaries can provide opportunities to expand our perspectives on the issue of climate change.

What is a worldview?

A worldview is a conception of the world. Worldviews are made up of a combination of beliefs, assumptions, attitudes, values, and ideas that together form a comprehensive model of reality (Schlitz, Vieten, and Miller 2010). Worldviews can be also described as "overarching systems of meaning and meaning making that to a substantial extent inform how humans interpret, enact, and co-create reality" (Hedlund-de Witt 2013: 156). Worldviews represent shared understandings about how society functions, including the role and purpose of social interactions and the way that people should be governed. They both influence and reflect what people believe and what they value, i.e., what they consider worth preserving, protecting, or nurturing. They also influence how we think about the past, present, and future and the power of the individual to control his or her own destiny. For example, some worldviews may see the future as predetermined, whether by divine providence, karma, or biology, leaving little room for human agency. Others may view the future as yet to be decided, believing that humans, themselves, play a key role in shaping what happens. Still other worldviews share the philosophical idea that the future does not exist, with the only real moment being now.

Every worldview represents a perspective or a view from a certain vantage point. Although worldviews reflect what appears to be true from a particular perspective, they are always partial, focusing attention on some things and not on others. Each worldview has blind spots representing conditions, characteristics, or relationships that are either not seen or not considered meaningful or relevant. People are often unaware of their own worldview and may seldom consider the worldviews of others. Yet being aware of worldviews can help to make sense of different perspectives on complex issues such as climate change and to appreciate the diversity of relationships between people and the planet. While worldviews represent what is "real" to an individual, community, group, or culture, they are not equivalent when it comes to caring for other people and the planet.

Categorizing worldviews

Multiple worldviews coexist today, and they can be categorized and classified in different ways. A commonly used approach is to categorize major world-views as traditional, modern, and postmodern and to recognize emerging, integral worldviews (Wilber 2000; de Witt 2015). In considering these categories and how worldviews may develop over time, it is important to recognize that broad generalizations tend to obscure the complexity of culture and human development, often with the result of boxing individuals or groups into categories that may be an appropriate description at certain times and within some contexts, but not others. Nonetheless, these distinctions can be helpful, not only to help you track your own journey of learning and growing but also to make sense of the ideological conflicts that occur in society, including how disagreements over climate change may be rooted in different worldviews (de Witt 2015).

Traditional worldviews recognize truth as it has been prescribed or handed down through generations. They may, for example, emphasize a hierarchical organization of society, as well as the value of stability and security within communities. Challenges to absolute and accepted truths are often perceived as threatening to traditional worldviews. Scientific explanations for climate change may thus be rejected if they question traditional understandings of human–environment relationships. Religious leaders and community members who are respected or in positions of power or influence may be more effective messengers for climate change communication than scientific reports and news stories, particularly if they relate climate change directly to the values of a community. A recent example is the widely disseminated papal encyclical on climate change (Pope Francis 2015: 20), which stated unequivocally that: "Climate change is a global problem with grave implications: environmental, social, economic, political and for the distribution of goods. It represents one of the principal challenges facing humanity in our day." This message from Pope Francis may be taken more seriously by those with traditional worldviews than if the same message were expressed by environmental scientists or social activists.

Modern worldviews value freedom, progress, and achievement. Modern worldviews stress individuality and the importance of rational inquiry – the search for a truth that is not declared or given, but discovered. The scientific method associated with the Enlightenment has led to many important breakthroughs, and new discoveries are continually challenging what was previously understood and accepted as "fact." For example, over the past centuries biology and geology have drawn attention to processes such as evolution and plate tectonics, and quantum mechanics has challenged atomistic, mechanistic, and deterministic worldviews. Science has considerable authority within the modern worldview, and climate change science and expert opinion are likely to be valued. While climate change may be viewed as a legitimate risk, modern worldviews place great trust in technological progress as a means of addressing it.

Postmodern worldviews recognize that there are many truths, and that these truths are relative to cultural and social contexts. Postmodern worldviews have a tendency to question positivist science and paradigms of economic growth and progress. There is an emphasis on pluralism, favoring values such as equity, inclusivity, and justice. While the idea of community is emphasized, it tends to be interpreted as fluid rather than fixed; non-traditional groups, subcultures, and identities are recognized and embraced by the postmodern worldview. Postmodern worldviews emphasize the social construction of knowledge, including the idea that truths are neither given nor discovered but created. Indeed, the very recognition that there are different discourses on climate change might be understood as reflecting a postmodern worldview. While this worldview deconstructs nature by denying the existence of one truth or reality, it also seeks to protect it from exploitation and destruction.

Integral worldviews describe holistic ways of seeing the world. Integral (or integrative) worldviews draw attention to the dynamic relationship between exterior or "objective" worlds and interior or "subjective" worlds. They can be considered a meta-perspective that draws together many paradigms, theories, fields of knowledge, and methodologies to provide a network of perspectives (Wilber 2000). Integral worldviews tend to be positive, inclusive, emancipatory, and reflexive, synthesizing multiple perspectives and ways of knowing and recognizing that all worldviews hold important truths (Hedlund-de Witt 2014). Integrating "sense and soul," they tend to value scientific rationality while appreciating spirituality as a unified means of seeing the world (Wilber 1999). They also tend to draw on developmental perspectives, recognizing that meaning making can evolve and expand over time to embrace greater complexity and include wider circles of care in relation to both humans and non-humans. It has been suggested that an integral worldview would be helpful in grasping the dynamism, the emancipatory imperative, and the full complexity of climate change, while also drawing attention to the deeper human dimensions of the issue (O'Brien and Hochachka 2010).

Changing worldviews

In considering these different categories of worldviews, it is important to keep in mind that the worldviews of individuals and groups are changing. Rather than being fixed or immutable, worldviews are continually updated, influenced, or activated by changing social contexts. For example, the World Values Survey, carried out in 80 countries every year since 1985, shows how values and worldviews are moving from traditional to modern, or from modern to postmodern, and in some cases even towards post-postmodern values (Inglehart and Welzel 2010).

Adults may experience multiple shifts in worldviews within a lifetime, as people are living longer and are exposed to diverse ideas and changing social conditions (Kegan and Lahey 2009). These shifts may also happen through the experience of personal crises, such as a significant health issue or loss of a family member. However, more often they occur when assumptions are challenged, new scientific discoveries are made, and innovative technologies are spread. They can also be catalyzed by major events that influence shared meanings and emotions. These may include something like the "Earthrise" photo from the Apollo 8 mission described earlier. Historical events or catastrophes, such as the September 11, 2001 attack in the United States or the 2011 Fukushima nuclear disaster in Japan, may also influence worldviews. However, every event can be interpreted through multiple lenses, such that they can trigger either progressive or regressive change. Sometimes worldviews are transformed through positive experiences, for example, through travel to new places, exposure to new ideas and perspectives, or practices such as meditation.

Worldview transformations tend to involve a significant change in everyday approaches to meaning making, which is sometimes described as a shift in mindsets or consciousness. Schlitz, Vieten, and Miller (2010: 21) discuss how social consciousness, or "conscious awareness of being part of an interrelated community of others" changes over time. This concept captures the level of explicit awareness that a person has of being part of a larger whole, and the extent to which she or he feels connected to others. Schlitz, Vieten, and Miller (2010) describe five nested levels of social consciousness: (1) embedded; (2) self-reflexive; (3) engaged; (4) collaborative; and (5) resonant (see Figure 4.2).

Recognition of the capacity to develop and transform over time to embrace more complexity and larger circles of care is critical to the integrative discourse on climate change (Hyman and Jalbert 2017). However, it cannot be assumed or taken for granted that worldviews are progressing towards universal, benevolent values motivated by **self-transcendence**. In times of uncertainty, instability, or crisis – when beliefs, values, and worldviews seem threatened – some people or groups may resonate with worldviews that make them feel more secure, even if it means limiting their circle of care to those within their own group or "tribe" (Schlitz, Vieten, and Miller 2010).

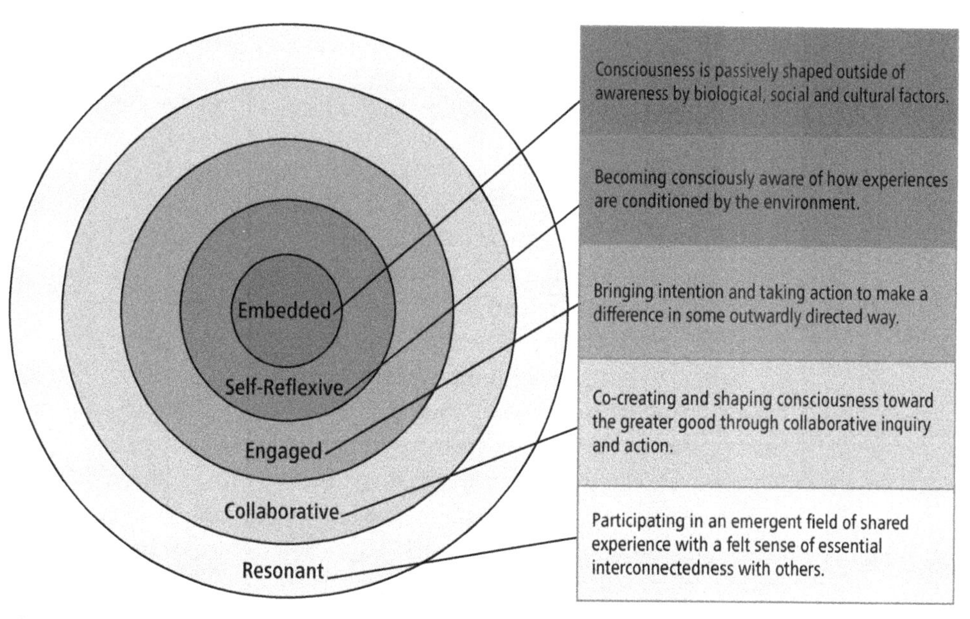

Figure 4.2 Five levels of social consciousness.
Source: Schlitz, Vieten, and Miller 2010

Views of nature and climate change

Worldviews reflect different philosophical understandings of the connections between humans and the environment. Some worldviews emphasize ecological interconnectedness or the idea that humans are caretakers or stewards of nature, while others are based on ideas of human exceptionalism, which is the view that humans are categorically different from all other species. Below, we discuss the connections between worldviews and views of nature, and show how they contribute to different perspectives on climate change.

Views of nature

There are many ways to view nature, and these views have changed significantly over time, influencing how people understand and relate to the abstract ideas of climate and climate change (Hulme 2009). One widely used system for categorizing views of nature comes from cultural theory, which suggests that there are four fundamental ways of depicting nature, each of which illustrates and reflects different worldviews or "ways of life" (see Figure 4.3) (O'Riordan and Jordan 1999). These "caricatures" or myths of nature include different assumptions about stability and different perceptions of the processes that affect stability, which in turn influence the policies that are considered appropriate for managing nature (Holling, Gunderson, and Ludwig 2002). Although there are other typologies for describing how humans see nature, these four views effectively illustrate how different understandings of nature can mediate human responses to environmental issues.

1. *Nature as capricious*: Nature is perceived as being a lottery that is out of the hands or control of humans. This fatalistic approach suggests that humans cannot influence Earth's natural systems and that natural events are the result of forces such as spirits, God's will, natural variation, or sheer randomness. Nature is beyond human control.
2. *Nature as benign*: Nature is perceived as highly resilient and cannot be permanently harmed by human actions. Instead, nature can bounce back from whatever type of stress humans place upon it. Nature has **instrumental value**; it exists to serve the needs of humans.
3. *Nature as tolerant*: Nature is perceived as able to withstand perturbations and "bounce back" from some degree of disruption. However, there are tipping points, where too much disruption can cause the system to move abruptly into a new state. If nature is treated with care, it can be managed and used for human purposes.
4. *Nature as fragile*: Nature is perceived as being sensitive and vulnerable

to disturbances. This view suggests that the Earth's natural cycles and processes can easily be disrupted by human activities and, once these occur, systems are not able to recover. Nature has intrinsic value and must be protected and treated with the utmost care.

While each of the categories described above reflects a different way of viewing nature, implicit in all is an understanding that nature is separate from humans. In fact, the very idea of a "view of nature" implies seeing nature as external to humans and to society. A number of researchers have pointed out that such rigid distinctions between humans and nature can lead to misplaced or shortsighted policies, which often contribute to the collapse of natural resources systems that humans rely upon (Holling, Gunderson, and Ludwig 2002). In their influential book *Panarchy*, Gunderson and Holling (2002) identify an emerging fifth view of nature, whereby it is considered dynamic, evolving, and adaptive. This view of nature recognizes characteristics such as chaos and order, self-organization, non-linearity, discontinuous

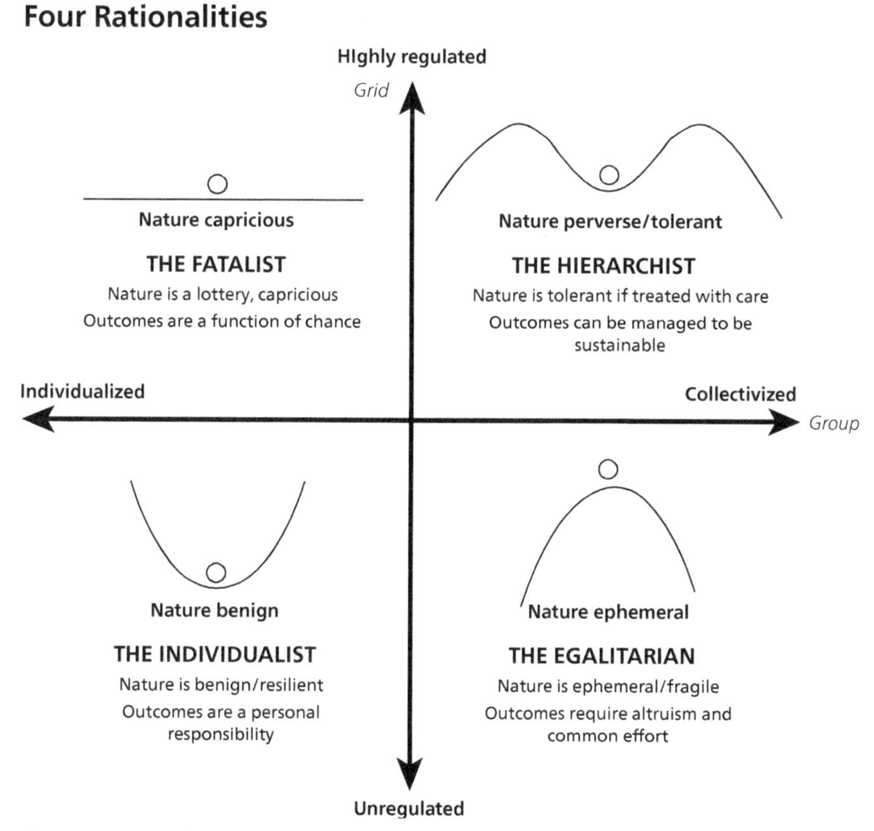

Figure 4.3 Myths of nature and cultural theory.
Source: Based on Schwarz and Thompson 1990 and O'Riordan and Jordan 1999

change, and complex behavior as components of adaptive, evolving socio-ecological systems.

As we discussed in the previous chapter, the separation of humans from nature is often referred to as a dualistic view. By contrast, **non-dualistic views** see humans and nature as interconnected, entwined, or entangled. Non-dualistic views range from acknowledging that human influences have already affected everything from soils and vegetation to the atmosphere, to recognizing that humans themselves are nature, in that 99% of the human body is made up of oxygen, carbon, hydrogen, nitrogen, calcium, and phosphorus (60% of the adult body is water), and the human body is host to trillions of bacterial cells. Emergent perspectives from the environmental humanities view humans not as separate, self-contained entities but as part of a larger interconnected system or as a collection of open-ended systems (Alaimo 2010). More generally, these perspectives are often critical of anthropocentrism (prioritizing humans) and recognize the agency of non-humans (Coole and Frost 2010).

Non-dualistic views of nature are characteristic of some cultures and religions. Non-duality, or "oneness," has long been a central spiritual teaching in Buddhist, Hindu, and Taoist philosophies. Non-dualistic worldviews see humans and nature as deeply integrated – as in "not-two" to begin with – thus they take a very different perspective on climate change. According to non-dualistic views, nature is recognized as arising within one's own awareness, and there is a realization of Nature beyond the vast diversity of natures. From this perspective, one gains a sense of profound interconnection between beings, whether they be humans or other species, or through culture, language, meaning, or consciousness.

Many **indigenous worldviews** conceive of nature as active, exhibiting agency and, to a certain extent, sentient or capable of perception (Heyd and Brooks 2009). Indigenous worldviews are often associated with **traditional ecological knowledge**. As defined by Berkes (2008: 7), traditional ecological knowledge is "a cumulative body of knowledge, practice, and belief, evolving by adaptive processes and handed down through generations through cultural transmission, about the relationship of living beings (including humans) with one another and with their environment." Core beliefs associated with traditional ecological knowledge include an ethic of non-dominant, respectful human–nature relationships, or what Berkes (2008) refers to as "a sacred ecology." These views are also reflected in deep ecology, which emphasizes harmony with nature, the intrinsic value of all species, and a questioning of the assumptions embedded within the dominant modern worldview (Naess and Rothenberg 1989; Devall and Sessions 2001).

Linking views of nature to understandings of climate change

The differing views of nature described above have a powerful influence on how we interpret and respond to the issue of climate change. While worldviews do not directly map on to the specific climate change discourses presented in chapter 3, the language or terminology of one discourse may speak more readily to one worldview than to others. The climate change discourse that we feel most "at home" with is likely the one that resonates most with our worldview. Among the dualistic views of nature mentioned above, each promotes a different way of seeing the connections between human activities and the climate system. A person who views nature as capricious and out of human control, for example, would likely consider climate change as something that cannot be caused by human activities. A person who views nature as benign might argue that the atmosphere and oceans can tolerate additional carbon dioxide from human activity. While this person might acknowledge that human actions can affect the climate, they are more likely to see nature as highly resilient and able to recover from any type of perturbation. Both of these views – nature as capricious and nature as benign – would suggest that there is little that can or should be done about climate change.

People who see nature as either highly fragile or manageable if treated with care are more likely to adhere to the **precautionary principle**, which emphasizes preventative action in situations of uncertainty and explores a wide range of alternatives to actions that are potentially harmful. The UNFCCC recognizes the precautionary principle and maintains that uncertainty is not a sufficient argument for delaying action. For those who view nature in a more instrumental way, protecting nature from climate change serves economic, cultural, political, or recreational interests. Acknowledgment of the benefits that the environment offers to human well-being is captured by the concept of **ecosystem services**. Examples of ecosystem services, which have been estimated to contribute more than twice as much to human well-being compared to global GDP (Costanza et al. 2014), include the provision of fresh water by glaciers, coastal protection by mangrove forests, absorption of runoff by green areas in cities, and pollination of agricultural crops by insects. Protecting and maintaining these ecosystem services thus offers a strong economic rationale for mitigating climate change. In contrast, a person who sees nature as having intrinsic value is likely to consider climate change as a moral and ethical issue (Gardiner 2004). Non-dualistic worldviews recognize that humans and nature are one, and that protecting the environment is protecting humanity.

In thinking about the linkages between climate change, worldviews, and views of nature, it is worth considering how individual and collective experiences of impacts might contribute to changing worldviews, both in the present and future. Experiencing an extreme weather event for the first time, whether a damaging hurricane or a destructive wildfire, may change the way that some people relate both to nature *and* society. In New Jersey, for example, some individuals who experienced damage to their homes during Hurricane Sandy in 2012 reported feeling a stronger connection and a more positive view of their community as a result of the support they experienced both during and after the hurricane (Solecki, Leichenko, and Eisenhauer 2017). Such direct experiences with climate change may trigger reflections on the role of neighbors and community and on the relationships between humans and the environment. Rethinking connections with neighbors, communities, non-human species, and the future can lead to a new awareness of one's own vulnerability and resilience, as well as of one's potential to influence the future.

Beliefs, identities, values, and emotions

Worldviews and views of nature do not come from nowhere. Nor do they simply represent the same world seen from different views. Instead, they create, enact, and perpetuate different realities. Reality is perceived, interpreted, and experienced from a multitude of physical, social, cultural, psychological, and spiritual perspectives. This means that climate change is not a "black and white" subject that can be understood in a right or wrong way. In fact, research on the relationship between climate change beliefs, values, attitudes, motivations, risk perceptions, and behaviors in the United States shows that adults respond to climate change in a variety of ways. The study, "Climate Change in the American Mind," which tracks US public opinion on climate change, consistently finds that most people (75%) fall somewhere in between the categories of alarmed and dismissive and are better described as being concerned, cautious, disengaged, or doubtful (Leiserowitz et al. 2015). Recognition that people relate to climate change in many different ways suggests a need to look more closely at the role of beliefs, values, social norms, and emotions in shaping and reinforcing worldviews and in activating and engaging people with climate change.

Beliefs and identities

Do you believe in climate change? This simple question has received an enormous amount of attention in research and in the media. National surveys have been conducted in many countries to measure the percentage of the population that believes that climate change is occurring and is caused by humans. Embedded in these inquiries is the assumption that if people believe in climate change, they will do something about it, either by changing their behavior or supporting climate change policies (Hornsey et al. 2016).

The question of belief in climate change is not just a question about whether one thinks the physical phenomenon is true or false. It is also related more broadly to belief systems. A belief can be defined as trust, faith, confidence, or acceptance that something is true. Beliefs relate to worldviews, in that "[b]eliefs constitute one of the ways that we describe the world we live in" (Nilsson 2014: 7). Countless beliefs, including beliefs about causality, self-efficacy (whether we can accomplish a task), the meaning of life, and so on, contribute to an individual's or group's worldviews. These beliefs are not random but are socially and culturally conditioned through personal and collective experiences, familial relationships, political discourses, religious teachings, formal education, or media messaging. Beliefs support mental models and understandings of systems, events, and experiences. They may be scientific (based on rational logic and evidence), common sense (based on experience), or learned (based on teachings or indoctrination). Beliefs not only shape our perceptions of and attitudes towards ourselves and others but also govern how much we think we can influence the present and the future.

Beliefs are closely linked to identities, i.e., the personal and social roles that people take on in different situations or at different times in their lives. A scientist, for example, may have a strong expert identity, which often privileges objective, measurable knowledge over tacit or experiential knowledge. Environmentalists campaigning for climate action sometimes have heroic identities associated with service and "saving the planet." Those who are skeptical of the evidence of climate change may sometimes have a strong identity as the critic or maverick, where pride is taken in not trusting evidence or in playing the role of "devil's advocate."

Beliefs also tend to be self-reinforcing, as people tend to pay more attention to facts that are consistent with their worldview and to dismiss those that are inconsistent. Indeed, research has shown that people tend to see what they believe (Lorenzoni and Hulme 2009); for example, people who believe that climate change is a problem are likely to see melting glaciers as a sign of climate change, whereas those who do not may invoke other explanations. Research from climate psychology has identified many ways to explain the differing public responses to climate change. For example, Stoknes (2015) discusses barriers that describe how the human psyche defends itself against scientific facts. Among these are cognitive distance and cognitive dissonance. Cognitive distance is the perception that climate change is socially, temporally, and geographically separate from our lives; cognitive dissonance is the state of holding thoughts, beliefs, or attitudes that are inconsistent with one another. With both distance and dissonance, the mind tends to favor beliefs that resonate with immediate concerns and priorities.

The notion that beliefs tend to be self-reinforcing is also reflected in talk of "echo chambers" and "living in media bubbles," whereby individuals only hear or read news and opinions that resonate with their own beliefs and worldviews. The next time you hear someone say "I believe in climate change" or "He doesn't believe in climate change," you may want to inquire about their views regarding the relationship between humans and nature, the role of social power and authority, and humanity's capacity to shape the future. To consider a problem from someone else's perspective can be challenging, yet it can also provide insights into what is important to them.

Values and social norms

Beliefs and identifies are also closely linked to values. In everyday usage, values may refer to desires, interests, needs, goals, attractions, or what is considered to have meaning based on individual and shared beliefs. More specifically, values describe what is intrinsically desirable to an individual or group and are often linked to underlying motivations, such as conservation, self-enhancement, openness to change, and self-transcendence (Schwartz 2007) (see Figure 4.4). The same values can be expressed in many ways, depending on the cultural and social context. For example, some may express security values through support for community-based adaptation programs and others by building walls and barriers to prevent climate migration. Values shared among a group of people, sometimes termed **cultural values**, have been found to play a key role in beliefs and attitudes about climate change (Kahan et al. 2012). If climate change is perceived as a threat to a group's values, information about it may be ignored or resisted. For example, if policies imply changes in environmental regulations, climate change is likely to be dismissed as a problem by those who place high value on individualism and freedom from government-imposed regulations (Leiserowitz 2006).

Another key influence on beliefs and values is social norms. Social norms refer to informal, shared understandings or agreements that govern the behaviors and actions of members of a group or society. Norms play an important role in the social organization of societies, and they are often internalized at an early age and perceived as if there were no alternatives – as if they were a given. As Schlitz, Vieten, and Miller (2010:24) point out: "the majority of people tend to naturalize social forces, unaware of their construction by political, economic, and cultural interests." Social norms are continuously created and maintained through narratives and practices, reinforced through rewards and feedback about what is right or wrong, and enforced through sanctions or punishments.

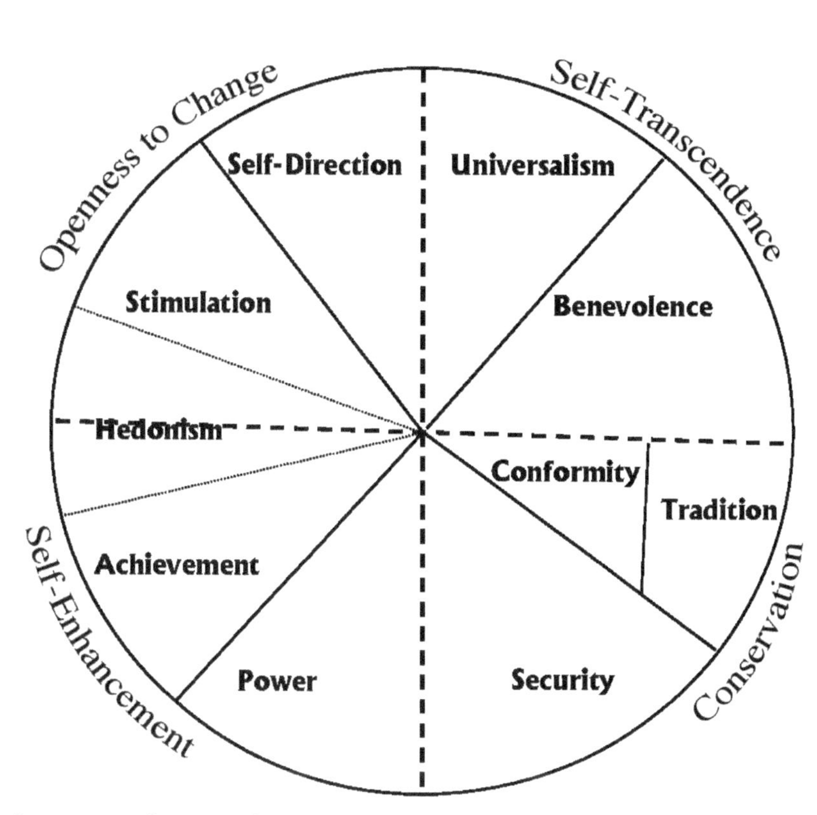

Figure 4.4 Schwartz's theory of basic human values.
Source: Schwartz 2007

Although social norms may appear rigid and robust, they are not fixed and can sometimes change quite rapidly. For example, in many countries, behaviors such as smoking in offices and littering were considered "normal" just a few decades ago but are now considered unacceptable and are often illegal (Nyborg et al. 2016). Emerging or new social norms tend to be reinforced when they are formally institutionalized through rules, laws, and regulations. Efforts to enact an international "ecocide law," for example, represent an effort to transform the protection of ecological systems from a social norm into a legal responsibility (Higgins, Short, and South 2013).

People seldom question or reflect on the core beliefs, values, and social norms that influence their worldview. Fixed beliefs often contribute to judgments and conflicts with others who do not share the same beliefs. When beliefs and assumptions are recognized as being conditional rather than unquestionably true, people are more likely to seek out new information and to consider a wider range of theories and explanations (Fazey 2010). We will return to the idea of transforming beliefs, values, and social norms in chapter 10. The point to keep in mind here is that viewing issues and objects from different perspectives can lead to new ways of seeing both problems and solutions.

Emotions

While beliefs, identities, values, and norms influence how we cognitively engage with climate change, our connection to the issue is also emotional. Emotions play a powerful role in our everyday lives, both consciously and unconsciously. They influence our perceptions of the world and how we process, interpret, and give meaning to events, places, circumstances, and activities (Ryan 2016). Many emotions are desirable and actively sought after (for example, pleasure, happiness, affection, gratitude, hope, love) and others are often considered negative and may be deliberately avoided (for example, fear, anger, despair, boredom, guilt). The power of emotions in evoking human responses has long been recognized by politicians, religious leaders, writers, artists, advertisers, and others, and there are many strategies to elicit emotional responses to achieve a desirable outcome or to avoid an undesirable one.

Emotions can serve as internal or individual barriers to processing or reacting to climate change information, as well as in motivating action (Salama and Aboukoura 2018). Climate change can bring up powerful emotional responses, such as fear, grief, uncertainty, inadequacy, disappointment, despondency, and despair (Fritze et al. 2008). Lesley Head (2016: 21) argues that "grief and climate change are inextricably entwined." Drawing from feminist theory, she criticizes a tendency to consider emotions as inferior to thought and reason and points out that climate change challenges this and other foundations of modernity. Research shows that many people respond to feelings, or what psychologists refer to as affect, more than they respond to facts, figures, and arguments that appeal to thoughts or cognition (Weber 2010). In describing the emergence of a "post-truth" era, D'Ancona (2017: 31) points to the growing influence of emotional narratives: "[I]n the new setting of digitalization and global interconnectedness, emotion is reclaiming its primacy and truth is in retreat." Affective responses are quicker and more automatic, and they are often linked to personal experiences, whether real or imagined (such as through film, literature, or art, as discussed below). Although emotionally engaging information can increase public attention to climate change, there is a risk of unintended consequences, including a decrease in concern about other risks because people have only so much capacity for worry (Weber 2010).

We have expectations for how the climate "is" and how it "should be" and may feel disoriented or a sense of loss when the climate is not what we expected. Glenn Albrecht has coined the term "**solastalgia**" to describe "the distress that is produced by environmental change impacting on people while they are directly connected to their home environment" (Albrecht et al. 2007: S95). Warmer than normal temperatures, less snow and ice, changes in the presence or distribution of species, and unfamiliar weather may trigger both a sense of solastalgia and grief (Head 2016). While climate change is often associated with negative emotions, in some locations warmer temperatures may be welcomed, leading people to conclude that climate change is beneficial. In such cases, it is easy to forget the systemic impacts of climate change, and its influence on species and ecosystems, extreme events, other places, and future generations.

The powerful influence of emotions such as fear and hope in generating understanding and action on climate change has been recognized by those working on science communication. Some argue that messages of "doom and gloom" about climate change should be avoided as they may depress people and make them feel despondent about the future. Others are convinced that an honest presentation of the situation is needed to stimulate urgent action. However, targeting fear or hope to leverage a specific action or outcome has been criticized as a misplaced strategy, as it overlooks the multidimensional qualities of emotional experiences (Chapman, Lickel, and Markowitz 2017). Research shows that emotional responses to messages about climate change are influenced by the beliefs, worldviews, and existing emotional states of individuals, and that communication strategies should connect to people authentically, rather than be used as a means of social engineering (Chapman, Lickel, and Markowitz. 2017). Indeed, as we show in the next section, artistic efforts to create authentic connections with climate change often tap into both cognitive and emotional dimensions of the issue. Such interventions highlight the "transformational potential" of emotion as a motivator for climate action (Ryan 2016: 6).

Connecting with climate change through film, fiction, and art

Artistic interventions provide an alternative way to the think about and connect with the issue of climate change (Galafassi et al. 2018). Creative practices can stimulate new processes of inquiry, political engagement, and imagination (Gabrys and Yusoff 2012). They may activate those who may not be enthusiastic about data, graphs, and scientific reports and also inspire scientists to think beyond the boundaries of their fields or disciplines. Artistic modes of communication take many forms: paintings, fiction, poetry, film, performance art, dance, and even comedy and music videos. Whether through stories, images, sounds, or other types of artistic expression, art can enable people to see the world from a different perspective. Art can also provide a direct link to emotions and introduce new ways of relating to the past, present, and future. In exploring examples of climate change art, we consider how the work of filmmakers, writers, and other artists has the potential to open up new possibilities and perspectives on humanity's capacity to respond to change.

Imagining the future through TV, film, and fiction

Climate change is becoming an increasingly visible topic in film, television, and fiction. Often these productions are intended to present a warning about the risks of climate change and the need to take action, or they provide a social commentary on humanity's seeming obliviousness to the dangers of a changing climate (Svoboda 2016) (see Box 4.1). For example, in the 2004 science-fiction disaster film *The Day After Tomorrow*, the lead character is a polar scientist whose predictions about the dangers of climate change are not taken seriously. The film's interpretation of the physical processes of climate change does not accurately reflect scientific knowledge (it suggested that climate change might alter the Gulf Stream ocean currents and almost instantaneously bring on a new ice age). Nonetheless, the film's compelling story and imagery succeeded in drawing widespread attention to the issue of climate change. The film also conveyed a not-so-subtle message about the connection between climate change, politics, and national security. During a memorable scene late in the film, climate-displaced residents of the United States seek and are provided refuge in Mexico. Some science-fiction films, such as *Interstellar*, present a world where climate change is destroying or has destroyed the conditions for sustaining life on earth. Other films like *The Martian*, *Gravity*, and *Avatar* present a "cosmic" view, i.e., a perspective that helps us relate to the universe as a whole and to see the Earth as part of a larger system. This shift in perspective can be a powerful incentive for taking better care of the Earth, which remains, for now, the only viable home for humanity.

Along with growing numbers of TV shows and films, novels and stories about climate change have become a recognized genre, sometimes referred to as **climate fiction (cli-fi)**. Many well-known, popular, and literary authors such as Kim Stanley Robinson, Margaret Atwood, and Ian McEwan have contributed to this genre, and the number is growing. According to political scientist Andrew Dobson (2010), a common ingredient of climate fiction is a grim future that is different from the present. For example, if a novel is set in the future, there are often flashbacks to a familiar past, and the characters in the novel confront the ethical and moral difficulties of these changing circumstances. Exploration of climate futures through fiction can provide a means to present different scenarios for a future climate and to consider the implication of our actions today through the use of imagination (Nikoleris,

Box 4.1 Imagining climate futures: Looking forward and back through film

Documentary films, such as *An Inconvenient Truth*, *The Last Flood*, *Cowspiracy*, and *Plastic Planet*, provide another form of narrative about climate change. *The Age of Stupid*, a film that is set in 2050, blends fictional and documentary images. It begins with fictional footage conveying images of mass ecological and social devastation. The audience learns that human civilization has been destroyed as the result of climate change. The film's main character is an archivist who is responsible for maintaining digital records of humanity's history, literature, and artistic achievements. The archivist spends his time examining documentary film clips from a period just before mass climate disruption and the unraveling of civilization. The film clips are supposed to be from the audience's "present day." They were filmed using documentary film clips from the early 2000s and feature real people who are experiencing early indicators of climate change, such as melting of glaciers in the Swiss Alps and unprecedented extreme weather events, but who feel that they have a very limited ability to effect change. For example, one individual featured in one of the film clips decides to stop traveling on airplanes in order to reduce his own greenhouse gas emissions. However, when he attempts to promote the use of wind power in his local community, he is unsuccessful, due to objections from neighbors.

The central question that the archivist asks throughout the film is why humans did not take action on climate change when there was still enough time to avoid catastrophe. The documentary clips provide a mechanism that pushes the audience to ponder this question as well. One of the interesting themes it brings up is "how we will be remembered" by future generations.

Think of an example of a film, book, or graphic novel that portrays a future of ecological and social devastation. How does the work explain what caused this dystopic future? Do you think the work conveys an effective message about the need and potential for taking action to prevent such a future?

Stripple, and Tenngart 2017). Literature has the potential not only to expand our conceptualization of solutions but also to humanize climate change, transforming data points into something that we can feel, taste, smell, and think about in a more personal way (Milkoreit 2016).

Much of climate fiction highlights dystopic futures under climate change and presents the struggles of "humans against nature" (Kaplan 2015). The literature showing positive futures includes classics such as *Ecotopia*, a 1975 utopian novel by Ernest Callenbach that helped to inspire the environmental movement. More recently, novels and short stories are starting to highlight connectedness and the possibility of creating alternative economies and societies. One example is "Entanglement," a short story by Vandana Singh about the connections among five people who are working seemingly independently in different parts of the world to address climate change.

Climate change has also become increasingly visible in popular culture. For example, the TV series *Game of Thrones* is considered by many political commentators and audience members to be a statement about climate change. Manjana Milkoreit (2017a) argues that this popular series has been used intentionally and instrumentally to pursue political goals, introducing a particular type of meaning making to the show's audience to contribute to better understandings of climate change and the politics surrounding it. Popular culture can provide a productive avenue for exploring value conflicts that are visible in the world today, whether related to drilling for oil in the Arctic, immigration policies for people displaced from their homes, or protection of biodiversity. The work can also raise political questions: Whose values count? Who decides the future? And how do we organize society in a way that is beneficial to everyone, rather than for a select few?

Climate change in visual and participatory arts

There is a saying that a picture is worth a thousand words, and pictures of polar bears standing on melting icebergs or wilting crops on parched soil have become iconic visual images intended to convey the meaning or implications of climate change. Photographs of coastal storms or flooding are also frequently used to highlight the devastation caused by a single event or to illustrate the impacts and potential future threats of climate change. Studying these iconic images, researcher Saffron O'Neill (2018) points out that they are inherently ideological: specific types of images can be used to naturalize powerful interests. Her research emphasizes the importance of connecting with people's socio-demographic and political affiliations and avoiding the use of cliché images.

Whereas journalistic or scientific photos are typically intended to document real physical events or the human consequences of climate change, artistic photos and paintings of climate change are deliberately created to convey an artists' personal interpretation of the issue. They are often constructed to provoke reflection and possibly evoke an emotional reaction. For example, *The Politics of Snow* series by artist Diane Burko compares photographs of glaciers from the last century to photographs taken by the artist today (see Figure 4.5). Using these photographs for guidance on how the physical dimensions of the glaciers have changed, Burko creates single paintings that show the same glaciers at different time periods. Her paintings not only convey information about the shrinking of the glaciers over time but also capture their beauty and evoke a sense of loss and sadness, as well as reflecting on the meaning of "the politics of snow."

In addition to photographs and paintings, artistic portrayals of climate change can also take the form of installation art. These site-specific portrayals can help to visualize what projections mean and to provoke questions about why climate change is happening. For example, a series of metal signs placed along a beach can powerfully convey projections of sea rise over future decades (see Figure 4.6). Another example of a large-scale installation art is Nele Azebedo's *Melting Men*, which are ice sculptures in the form of humans that slowly melt away over time, raising awareness of the fate of the Greenland and Antarctic ice sheets.

Participatory art, which entails direct engagement with the audience in the creative process, is often designed both to draw attention to the issue but also to inspire and motivate personal change and political action. For example, the website and group of artists behind Dear Climate (2018) seek to motivate individuals to think about their relationship to the climate

Figure 4.5 The Politics of Snow.
Source: Grinnell North Moraine, 1,2 (1922 and 2010) by Diane Burko, 2010. Collection of The Frances Young Tang Teaching Museum and Art Gallery at Skidmore College, gift of Michael Basta.

Figure 4.6 Rising sea levels: One prediction of where rising sea levels will end up at Cottesloe Beach, Perth, Western Australia.
Source: Photograph by go_greener_oz, Flickr, 2008

through distribution of provocative agitprop posters and podcasts and by encouraging users to write letters to the climate (see Figure 4.7). Participatory climate change art may take many other forms such as dance performance, sculpture, or photography.

As these various examples illustrate, artistic representations have the potential to provide powerful entry points for reflecting on how we have organized society and how our actions collectively influence the planet and ourselves. Imagining climate futures through film, fiction, and other forms of artistic expression is not only a way to explore what different climate futures will look like but can also open new ways of thinking about the problem today. In 2015, a global festival of artistic activity called ArtCOP was held in the run-up to the UNFCCC Conference of Parties (COP21) in Paris (van Renssen 2017).

While art can be a highly effective tool for engaging with climate change, we should not assume that art is a "magic bullet" that can effortlessly provoke positive shifts in consciousness and action on climate change (Miles 2010). Art is interpreted subjectively and can have different meanings for different people. Film, fiction, and art may sometimes reinforce values rather than challenge them. They can also create cognitive distance from the issue of climate change or come across as a form of political propaganda. Novelist and essayist Amitav Ghosh, author of *The Great Derangement: Climate Change and the Unthinkable*, suggests that the critique of climate change expressed

Figure 4.7 Dear Climate posters.
Source: http://www.dearclimate.net

through the arts may actually be a form of collusion, in that dissatisfaction has itself become another commodity and thus part of the current system that is contributing to "the great derangement" (Ghosh 2016).

Summary

The way we view the world and what we believe about ourselves and the environment have vital implications for how we perceive and approach climate change. In exploring the role of worldviews, values, beliefs, and emotions, it becomes clear that climate change is much more than simply an idea that one does or does not believe in. Rather, it is an issue that forces each of us to reflect on personal and shared assumptions and to consider new ways of seeing the world. Acknowledging and understanding differing worldviews can be a powerful way to make sense of different approaches and responses to the complex challenges of climate change.

Reflection questions

1 Consider your own view of nature. Can you recall a specific event or experience that influenced or changed your view of nature?
2 How do you think a person's worldview and values influence the climate change discourse that they feel most comfortable with?
3 Do you think that art, film, and/or fiction are effective for engaging people with the issue of climate change? Are there any downsides of this method of engagement?

Further reading

Ghosh, A., 2016. *The Great Derangement: Climate Change and the Unthinkable*. Chicago, IL: University of Chicago Press.

Hedlund-de Witt, A., 2013. Worldviews and their significance for the global sustainable development debate. *Environmental Ethics* 35(2): 133–62.

Leiserowitz, A., 2006. Climate change risk perception and policy preferences: The role of affect, imagery, and values. *Climatic Change* 77(1–2): 45–72.

O'Riordan, T. and Jordan, A., 1999. Institutions, climate change and cultural theory: Towards a common analytical framework. *Global Environmental Change* 9(2): 81–93.

Schlitz, M. M., Vieten, C., and Miller, E. M., 2010. Worldview transformation and the development of social consciousness. *Journal of Consciousness Studies* 17(7–8): 18–36.

5 THE SOCIAL DRIVERS OF GREENHOUSE GAS EMISSIONS

The elephant in the room

Many have heard the story about the blind men and the elephant. Six men who have never encountered an elephant each touch a different part of the animal – the trunk, the ears, the tusks, the side, the tail, the leg. Each comes to a different conclusion about what they are touching, based on their perception and experience. For example, the man who touches the trunk thinks he is touching a snake, while the man who touches a tusk is convinced that it is a spear. This ancient parable points to the human tendency to interpret personal experiences and perceptions as the whole truth. In relation to climate change, it is very easy to define the whole problem based on only one perspective. As we discussed in chapter 3, some discourses view climate change as an environmental problem, while others see it as a problem linked to capitalism or globalization, or perceive it as no problem at all. By focusing on only one or two dimensions of climate change, it is easy to miss the elephant in the room.

This chapter presents an integrative perspective on the drivers of climate change. Looking at the problem through multiple lenses, we examine where greenhouse emissions are coming from and consider the social processes behind them. We begin by presenting empirical patterns of CO_2 emissions by country and discussing how they reflect historical and contemporary patterns of industrialization. We then explore emissions through a number of different but interrelated lenses, including development and globalization, consumption, urbanization, and land use. We show that each of these lenses not only provides different insights on emissions sources and drivers but also widens the solution space for reducing climate risks through climate change mitigation.

National emissions patterns

Climate change is often described as a global challenge where all nations have an interest and stake in protecting shared resources and avoiding dangerous climate change. From a global perspective, a "Spaceship Earth" metaphor is often used to emphasize that "we are all in this together." A global view on emissions has been vital for understanding *how* the climate system is affected by rising emissions. However, this global view also hides profound differences in the "who, what, where, when, and why" of rising emissions. The roles that geography, history, politics, economics, and culture play in rising emissions become clearer when we consider other scales of analysis. We start by looking at national emissions patterns and how they are linked to development trends.

National emissions

A national lens on emissions is important for several reasons. First, nations negotiate and are signatories on international agreements on climate change, including the United Nations Framework Convention on Climate Change (UNFCCC) and the Paris Agreement. They are the primary units for emissions accounting, reporting, monitoring, and evaluation in connection with these agreements. National emissions also provide a quantitative metric for comparison of emissions among nations and over time. They highlight the high and low emitters, and provide baseline information for evaluating emissions-reduction efforts. These data can also be used to present a clear message about the carbon budget and the emissions reductions needed to limit warming to below 2°C (Global Carbon Project 2017).

National emission patterns can be documented and visualized in many different ways. Figure 5.1 presents data on annual, **territorial emissions** of greenhouse gases based on annual kilotons of carbon dioxide (CO_2) emitted for each country from fossil fuel usage and other activities. The figure illustrates some clear and well-known patterns, namely that current emissions are highly uneven across countries and world regions. It should come as no surprise that the world's largest economies and most populous countries also have the highest greenhouse gas emissions. Such countries include the United States, China, India, Russia, Germany, and Japan. The figure also highlights differences in emissions by major world regions. For example, North America, Europe, and East Asia stand out as high-emitting regions, whereas sub-Saharan Africa is notable for its lower emissions.

Examination of national CO_2 emissions per capita (see Figure 5.2) provides information on each country's emissions relative to population size. Mapping carbon emissions per person reveals dramatic differences across countries.

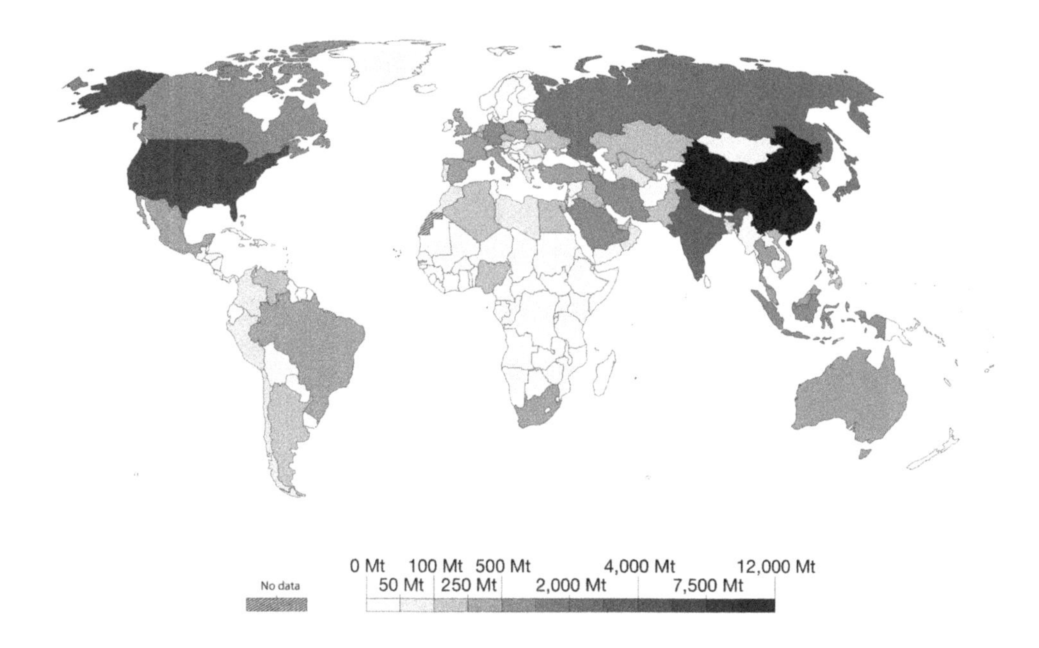

Figure 5.1 Annual CO_2 emissions by country, 2016, in million tonnes (Mt).
Source: Ritchie and Roser 2018

For example, average per capita emissions in the United States are roughly ten times the average emissions in India. By controlling for a country's population size, per capita data also confirm the conventional wisdom that the wealthier and more industrialized countries of North America and Europe, as well as Japan and Australia, have substantially higher per capita emissions than less industrialized countries. Other high emitters include oil-producing nations such as Oman, Bahrain, Kuwait, United Arab Emirates, Trinidad and Tobago, and Saudi Arabia. Yet Figure 5.2 also reveals that some highly industrialized countries have more modest per capita emissions profiles (e.g., China, South Africa, and many European countries). Most of the countries of sub-Saharan Africa, South America, and South and Southeast Asia fall into the low and very low emissions categories on a per capita basis.

Whether measured in aggregate or per capita terms, national emissions data confirm that there are stark and substantial differences in emissions by nation. However, present-day emissions tell only part of the story. Another way to think about national distributions of greenhouse gas emissions is to consider historical patterns and cumulative total emissions. Because CO_2 stays in the atmosphere for centuries, those countries that industrialized earliest, including the United States, the United Kingdom, and other

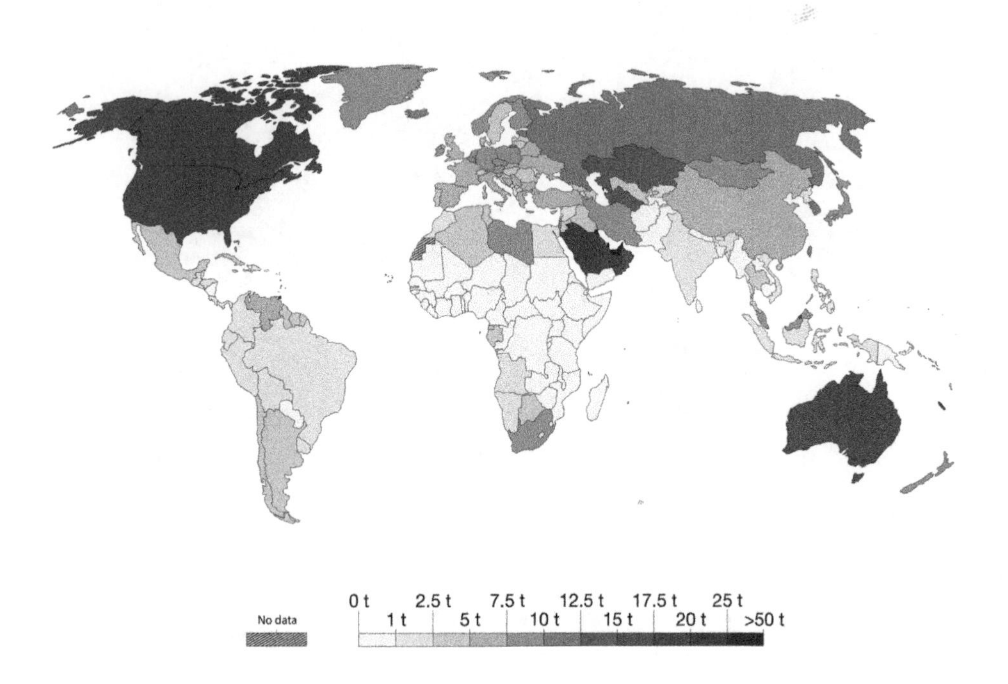

Figure 5.2 Annual CO_2 emissions per capita by country, 2016, in tonnes (t)
Source: Ritchie and Roser 2018

European countries, have the highest total emissions on a cumulative basis, even though emissions from these countries have been falling in recent years (see Box 5.1). China and India, two of the world's largest present-day emitters, have only become high emitters in recent decades (though much less so on a per capita basis). Many countries have contributed very little to cumulative total emissions. As we will discuss in chapters 7 and 8, those countries that have contributed least are often the most vulnerable to the impacts of climate change.

The question of how to count national emissions is not merely a matter of numbers; it also raises issues of fairness when it comes to responsibility for climate change mitigation. Should countries that only recently began to industrialize have the same obligations for emissions reductions as countries that industrialized in the past? Should low-emitting countries with petroleum resources, such as Myanmar and Mozambique, have the same opportunities to exploit oil and pursue industrial development as other countries had in the past? These issues are particularly relevant for international negotiations and have long been a point of contention between the Annex I Parties to the UNFCCC, which include industrialized (developed) countries and "economies

Box 5.1 Decoupling national growth from emissions

National emissions data reveal that some of the wealthiest countries have seen declining emissions in recent years, even as income levels have increased. The decline in emissions in many industrialized countries on both an absolute and per capita basis might suggest that economic growth can be decoupled from rising emissions. For less industrialized countries, the possibility of decoupling economic growth from emissions challenges the assumption that emissions reduction will necessarily constrain development, reduce standards of living, and interfere with poverty reduction. The evidence of decoupling also raises the possibility that future development opportunities might be realized without replicating the resource-extractive, high-pollution pathways that wealthier countries have taken. The premise of decoupling has, for example, been adopted by the government of Indonesia, which aims to reduce greenhouse gas emissions by 29% by 2030, while maintaining an economic growth rate of 7%. A study by Anderson et al. (2016) suggests, however, that the government's rhetoric has not been matched by a move away from exploitative and resource-intensive ecological practices such as palm oil production.

Observations that emissions tend to decline as income levels increase have led some scholars to suggest that national emissions follow the pattern of an environmental Kuznets curve, or EKC (Kaika and Zervas 2013a). The EKC model, which relates pollution levels to income, was first used to document the linkages between levels of development and environmental quality. The EKC depicts levels of income relative to overall pollution as an inverted U, based on observations that very poor countries and very wealthy countries typically have lower pollution levels than middle-income countries (see Figure 5.3). Applied to greenhouse gas emissions, the EKC has been used to support the argument that decoupling can occur with continued economic growth and development.

While the EKC might be intuitively appealing, particularly from the perspective of the dismissive discourse, the model has a number of limitations (Kaika and Zervas 2013b). In particular, it is more applicable to local pollutants that degrade air and water quality than to "invisible" pollutants such as CO_2 that do not remain in one location but spread globally. EKC approaches are also premised on the idea that each country is entirely responsible for its own emissions and does not take into account international trade between countries (Knight and Schor 2014).

Do you think that decoupling of economic growth and emissions is a feasible strategy for emissions reductions? Who might benefit from this approach and who might be harmed?

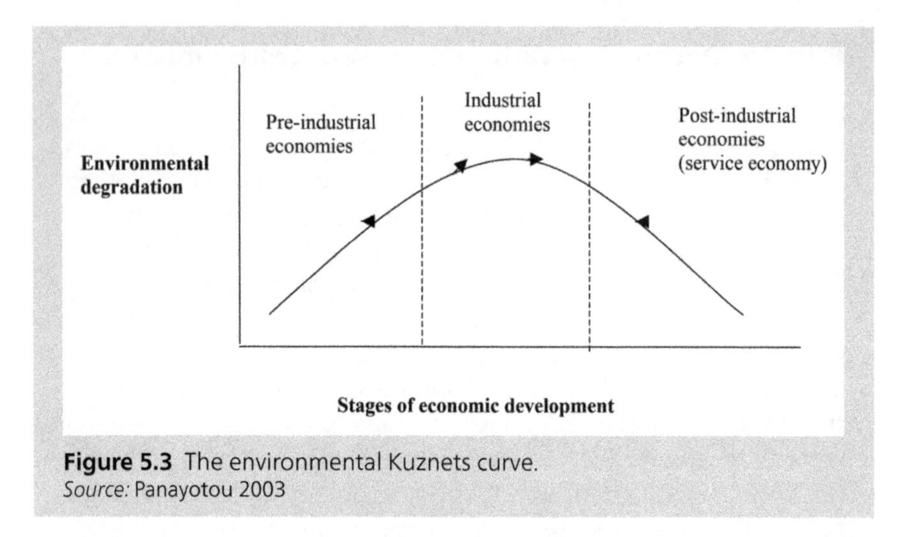

Figure 5.3 The environmental Kuznets curve.
Source: Panayotou 2003

in transition," and Non-Annex I Parties, which include low-income developing countries. The Paris Agreement of 2015 does not address historical emissions responsibility directly but instead acknowledges national differences as a starting point for future emissions reduction efforts. The agreement adopts a framework of nationally determined contributions (NDCs), whereby benchmarks for national emissions reductions are set through non-binding individual country pledges (Falkner 2016). The NDCs allow countries to commit to reductions that they think are feasible, and include flexibility to adjust pledges up or down.

National emissions and development

Each climate change discourse presents an interpretation of the reasons behind emissions differences across countries and draws attention to different opportunities for changing future emissions trajectories. Within the biophysical discourse, a widely used approach to explaining and projecting carbon dioxide emissions is an equation referred to as the Kaya identity. This quantitative approach is used to identify the key social and technological factors that influence greenhouse gas emissions in each country. The Kaya identity and related models are not simply a way to explain national emission patterns; they are also used to project emissions based on different future scenarios. As we discussed in chapter 2, emissions scenarios and projections are a key input in global climate models because the growth rate of emissions directly influences the rate and magnitude of future climate change.

The Kaya identity is derived from an environmental impact model known as I=PAT, which considers Impacts to be a product of Population, Technology, and Affluence (Rosa and Dietz 2012). Building from the I=PAT approach, it represents emissions as a multiplicative function of the following factors:

Emissions = Population * Affluence * Energy Intensity * Carbon Intensity

where:

Emissions = total emissions of carbon dioxide
Population = total number of persons
Affluence = gross domestic product per capita
Energy Intensity = energy use per unit of GDP
Carbon Intensity = carbon emissions per unit of energy.

An innovative contribution of the Kaya approach is how it treats population. Population growth is often singled out as the key driver of environmental degradation and emissions growth (Satterthwaite 2009). Big increases in global population occurred with the Great Acceleration during the latter half of the twentieth century, when mortality rates declined as the result of improved sanitation, more intensive agriculture, and advances in health care and medicine. As we discussed in earlier chapters, this was the same period during which greenhouse gas emissions took off. While population plays a role in emissions growth, it is important to remember that consumption, technology, and energy sources are often more influential (Kirby and O'Mahony 2018). The Kaya identity demonstrates how affluent countries with relatively small populations may have greater emissions than relatively poor countries with larger populations. For example, Norway's population of 5.3 million is roughly half the population of Burundi (10.5 million), yet Norway's emissions in 2014 were estimated to be more than a hundred times higher than those of Burundi (47,627 kt CO_2 for Norway versus 440 kt CO_2 for Burundi) (World Bank 2015). This approach also explains how smaller countries that are highly dependent on fossil fuel energy may have higher emissions than countries that are larger or more affluent yet have access to renewable energy resources.

Differences in energy intensity of production across countries can reflect many factors, including the overall structure of a country's economy. Many of the world's highly affluent countries have relatively small shares of their economy based on energy-intensive manufacturing industries and instead rely largely on "cleaner" postindustrial sectors such as financial services, health care, insurance, and education. Natural resource endowments also play a role, particularly when it comes to the development of hydropower, wind, geothermal, and solar energy. Many other factors may also influence future greenhouse gas emissions. For example, in examining Chinese emissions projections to 2030 based on the Kaya identity, Grubb et al. (2015: S26) note that "different assumptions around possible structural changes in the balance of the Chinese economy would have quite different impacts on the trajectory of future Chinese energy demand, which in turn would affect the trajectory of Chinese emissions." Assumptions are thus critical to future projections, including, for example, the possibility that new climate policies will be implemented or that the costs of renewable energy will decrease to the point where there is a large-scale shift away from fossil fuels.

While the biophysical discourse looks to quantitative and systematic frameworks for explaining and projecting national and global emissions, the critical and integrative discourses reveal some of the limitations of the Kaya identity and similar approaches. The critical discourse reminds us that national development trajectories are not simply a given but have historical roots in colonial conquests, resource exploitation, and unequal terms of trade, among other things. Models and projections of national emissions offer little possibility of questioning the status quo of a highly unequal world and ignore social and political factors that perpetuate these inequalities. National emissions accounts also ignore power dynamics between countries, which may allow wealthy, industrialized countries to "outsource" their pollution and emissions associated with production and manufacturing to other locations.

The integrative discourse raises questions about how the Kaya identity and related models conceptualize the future. Within these models, the future is typically projected based on linear extrapolation of past trends, with little recognition of human agency and collective will (Kirby and O'Mahony 2018). The projections do not incorporate the possibility that changing mindsets, social norms, and institutions may dramatically alter future consumption practices, resource use, and emissions patterns. In short, the models do not offer opportunities to question assumptions and ways of thinking that reinforce a high-emissions status quo, and thus they may underestimate the potential for social change.

Globalization and the spatial displacement of emissions

Counting national emissions and explaining the variations by country seems fairly straightforward. As discussed above, high levels of national greenhouse gas emissions are generally seen as a necessary side effect of economic development and industrialization. Territorial emissions are higher in more industrialized countries because these countries have more energy-intensive infrastructure, transportation systems, and built environments. Industrialized countries also have higher levels of income and wealth, and large shares of the populations of these countries are able to afford energy-intensive products, services, and lifestyles. Such an explanation assumes a fairly linear relationship between industrial development and emissions (i.e., higher levels of development lead to higher emissions). However, explaining the reasons *why* emissions vary by country is actually more complicated, particularly when we take into account the influence of globalization and its implications for emissions accounting and responsibility.

Over the past several decades, the world economy has become increasingly globalized. Along with expansion of international trade and integration of production and financial systems, globalization also entails the diffusion and uptake of social and technological innovations, increased connectivity, and the spread of a culture of consumption (Sklair 2002; Dicken 2015). Although globalization is often seen as universal and all pervasive, globalization processes are highly uneven, particularly in terms of who benefits and who is harmed. Globalization is implicated in growing polarization of income and wages, uneven rates of economic growth, differential access to political power, and limited or partial diffusion of new technologies (Leichenko and O'Brien 2008). Globalization also has implications for the spatial distribution and growth of greenhouse gas emissions.

Globalized trade and production have contributed to a **spatial displacement of emissions** whereby countries are able to reduce their territorial emissions through import and consumption of products that are manufactured elsewhere. In 1970, the United States, Europe, and Japan were the leading producers of manufactured goods. Since then, the manufacturing locations of products, ranging from mobile phones, computers, and electronic appliances to clothes, toys, and motor vehicles have increasingly shifted to China and India, as well as other countries of East and South Asia and Latin America. During this same period, the economies of the United States and many western European countries have become increasingly oriented towards "cleaner" and lower-emission service and information-based sectors such as education, health care, and finance. Indeed, many of the manufactured products that are consumed within these countries are produced elsewhere. Because territorial greenhouse gas emissions are credited to countries of production, rather than to where resources or products are consumed, many of these service-oriented countries have experienced emissions reductions over time. In the case of the European Union, for example, territorial emissions fell by 6% from 1990 to 2008, which was close to the 8% target that it agreed to in the 1997 Kyoto Protocol. However, by some estimates this reduction was more than offset by emissions associated with consumption of the goods and services that the European Union imported from other countries (Peters et al. 2011).

In terms of emissions accounting, recent estimates of emissions associated with international trade, sometimes referred to as **consumption-based emissions**, suggest that 26% of global greenhouse gas emissions come from the production of goods for export (Peters et al. 2011). Recognizing the importance of international trade, proposals have been made to base national emissions inventories on the goods that are imported into a country as well as those goods produced for local markets (Davis and Caldeira 2010). Trade-based, national emissions calculations would include emissions from the production of all goods and services that are consumed within a particular country, including all imports, while subtracting emissions associated with exports (Knight and Schor 2014). Inclusion of international trade in emissions calculations could potentially address some of the concerns over fairness in emissions responsibilities between countries. Complications remain, however, because a large share of global trade happens within multinational firms that have production facilities spread across many countries. Moreover, emissions associated with international shipping are not typically included in national emissions counts (see Box 5.2).

While globalization was once described by Thomas Friedman (2005) as making the world "flatter" in terms of opportunities, globalization also means that a small number of large multinational firms exert substantial control over **global production chains** for products ranging from automobiles to computers to chocolate hazelnut spread (see Figure 5.4). With sourcing, production, marketing, and distribution occurring across many countries, it becomes difficult to identify how or where to attribute the carbon emissions

Box 5.2 Accounting for missing and hidden emissions

Emissions counting is complicated by emissions that are missing from greenhouse gas inventories. For example, data on air travel and shipping are typically excluded from national emission accounts (Afionis et al. 2017). Emissions associated with air transport and shipping together are estimated to account for roughly 6% of CO_2 emissions globally (Pearce 2016) but are difficult to assign to one country or another. Military arms production and related expenditures are another example of missing emissions. Countries such as the United States, France, Germany, and the United Kingdom are leaders in the US$1.8 trillion global arms trade, along with Russia and China. The contribution of national and global arms production and military activities to greenhouse gas emissions is thought to be substantial, yet not considered in national inventory calculations. While airline and shipping were excluded from the 2015 Paris Agreement, countries may elect to include emissions from military activities in their nationally determined contribution (NDC) (Neslen 2015).

How do you think emissions should be assigned for passenger air travel? Should the country of origin, the country of destination, the country where the airline or shipping company is based, or the home residency of passengers take responsibility for emissions?

Figure 5.4 Global value chain of Nutella.
Source: De Backer and Miroudot 2013. Based on Ferrero, source map and various online sources.

of particular products. Recognition of the influence of multinational firms over emissions patterns also draws attention to the outsize influence of a relatively small group of decision makers linked to the fossil fuel industry, as we discuss in chapter 6.

The culture of consumption

The emission lenses that we have considered so far may seem very distant from our everyday lives. As individuals, we often feel that there is little we can do to influence global and national emissions, international trade flows, and multinational corporations. Yet we also have a tangible connection to emissions through our everyday practices, especially those associated with high levels of consumption. Consumption-based explanations for rising emissions draw attention to individual and collective choices and societal decisions and expectations about how and where we live, what we eat, how we move around, and all of the privileges and conveniences that we often take for granted.

The rise of consumer culture

Consumerism is rapidly becoming a dominant ideology globally. While the emergence of consumer identities began as far back as the fifteenth century (Trentmann 2016), a remarkable increase in the pace of global consumption occurred during the Great Acceleration. Rising consumption during this period was connected to many factors, including technological innovation, the rebuilding of infrastructures and economies after World War II, responses to surplus production, and the influence of advertising and media in creating an insatiable demand for products and services (Ehrhardt-Martinez et al. 2015). The globalization of consumer culture has many dimensions, but fundamentally it entails the spread of mass consumption ideals and practices. This includes a vision of "the good life," which Leslie Sklair (2002) describes as an "ideology of consumption." Although achieving this "good life" is a reality for a relatively small portion of the population on a global basis (see Box 5.3), the ideology nonetheless influences consumer practices worldwide.

The proliferation of "fast food," "fast fashion," and "throwaway" items, as well as rising global demand for products ranging from automobiles to smartphones, are characteristic of the globalization of consumption. Nearly all of these products contribute to rising energy usage and accelerated resource consumption. Harold Wilhite (2016) describes the rising consumption of energy-intensive goods and services as the "**habits of capitalism**." He argues that the expansion of the capitalist economy and its pillars – economic growth, individual ownership, product differentiation, and product turnover – have contributed to high-carbon habits and lifestyles, including larger sizes of houses, energy-using technologies and appliances, greater expectations for thermal comfort, higher standards of personal hygiene, greater mobility, and refrigerator-dependent food habits. As noted by Princen (2003), this culture of consumption is closely linked to the capitalist logic of profit maximization and market efficiency rather than to other logics that might emphasize principles such as restraint, precaution, and sufficiency.

The ongoing digital revolution, which is dramatically changing the way people shop and consume many products, represents a new dimension of global consumption. Although online shopping may avoid a trip to the mall, the production, shipping, and home delivery of goods and products still use energy and resources. The products that allow us to communicate and remain connected – smartphones, computers, tablets – are energy intensive to produce and use, and because of their relatively short lifecycles, they contribute to a growing volume of electronic waste, or e-waste. The proliferation of e-books and online media that can be read on computers, tablets, or devices has led to a decline in paper use, and the streaming of films, videos, and music has decreased the demand for physical objects such as CDs and DVDs. However, the "cloud" of massive data storage units, which allow

Box 5.3 Carbon inequalities

Drawing attention to consumption brings out differences in emissions across groups and individuals. Regardless of country of origin, wealthier individuals typically have much higher emissions than less well-off individuals. The concept of carbon inequalities illustrates how emissions associated with consumption vary across income groups, both nationally and globally. Figure 5.5, which is based on data compiled by Oxfam (2015), presents the percentage of CO_2 emissions associated with lifestyle-related consumption by income group. The figure shows that the richer half of the world's population is responsible for roughly 90% of lifestyle and consumption-related emissions, whereas the poorer half is responsible for only about 10% of total global emissions. When the data are further sorted into income tiers, we see that the richest 10% globally are responsible for roughly 49% of consumption emissions. For those in the highest income tiers, lifestyle emissions stem largely from taken-for-granted conveniences of modern life, such as in-home electricity, automobiles, airplane travel, and access to and use of a wide variety of foods and consumer products, many of which are not available to those in the lowest tiers of income distribution.

Do you think that assignment of emissions responsibility based on income and wealth level is preferable to assignment by country? Do you think that individuals in the highest income tier have a greater responsibility for taking action to address climate change?

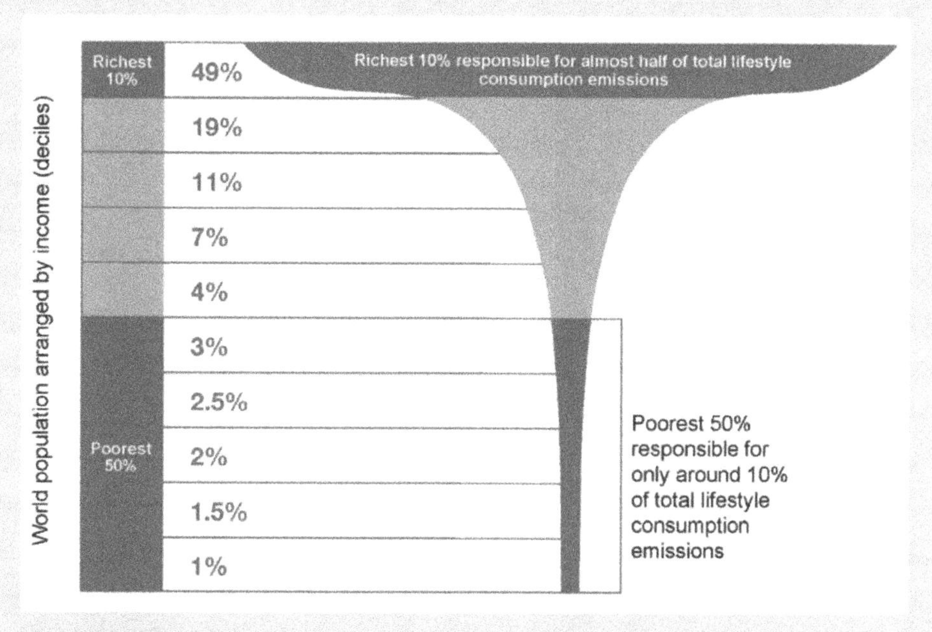

Figure 5.5 Percentage of CO_2 emissions by world population.
Source: Oxfam 2015

24-hour instant delivery of digital content to billions of users, is also highly energy intensive and is rapidly growing.

Carbon footprints and social practices

Emissions associated with consumption are often described through the concept of carbon footprints. **Carbon footprints** refer to the amount of carbon dioxide released into the atmosphere as a result of the activities of individuals, groups, organizations, cities, or other entities. They are typically represented as tons of CO_2 emitted on an annual basis and provide a way of estimating and comparing carbon emissions based on consumption habits. For individuals, carbon footprints are usually calculated by estimating carbon emissions associated with energy usage from all facets of daily life, including transportation and travel, food choices, clothing choices, home heating and cooling, electricity, and goods and services consumed. Carbon footprints can also be estimated for specific activities such as air travel. Many digital tools have been developed to help individuals estimate their overall carbon footprint and emissions associated with different activities (Pandey, Agrawal, and Pandey 2011).

Calculating individual carbon footprints not only helps to illuminate how consumption contributes to greenhouse gas emissions; it can also be used to provide guidance on areas where emissions reductions are possible. The links between personal consumption and emissions have received significant attention in climate policy, and they are widely seen to supplement or complement national strategies for emissions reductions. Communication about climate change and its causes often ends with an appeal to "what *you* can do." In most cases, these suggestions emphasize personal choices and behavioral changes. Shifting from using an automobile to using public or non-motorized transportation, minimizing or eliminating air travel, or changing diets are often presented as ways to reduce individual carbon footprints (Wynes and Nicholas 2017).

Viewed through a lens of individual behavior and consumer choice, which is emblematic of the biophysical discourse, carbon footprint calculations can imply that consumption decisions are entirely in the hands of the consumer. The choice between a gasoline versus an electric car, for example, is understood as a simple matter of preference, cost, and convenience. Perspectives from the critical discourse bring out structural factors that influence behaviors and choices. Social practice theory, for example, draws attention to the idea that behavioral choices do not reside solely in individuals but are socially produced and reinforced through norms, institutions, and infrastructure (Shove 2010). For example, some of the largest determinants of transportation options have to do with whether a person lives in a city, suburb, or rural area, how close they live to their workplace or school, and whether or not public transit is available. In areas with good public transit systems, communities and neighborhoods with better access to public transport are often more expensive; housing in these areas may be unaffordable for many who might otherwise wish to reduce automobile use.

Many other everyday practices are also constrained by factors that are outside the control of individuals (Hargreaves 2011). An individual may choose to use less energy at home by turning off lights, unplugging appliances, lowering inside temperatures in the winter, and minimizing use of air conditioning. Yet, unless one is building a new home, individuals often have limited say in the type of heating and cooling systems that are installed and even less ability to decide where their home energy comes from. Similarly, an individual can choose to reduce or minimize air travel for leisure or recreation, but air travel for some may be a vital aspect of their work. While reducing some air travel through video conferencing and virtual experiences is possible, the air travel associated with many consumer products is often hidden from carbon footprint estimates (see Box 5.2).

Equating concerns about climate change with personal carbon emissions and carbon footprints can potentially lead to feelings of guilt, hopelessness, and powerlessness. Ultimately, consumption choices and the degree to which individuals can reduce greenhouse gas emissions are influenced by larger social, economic, and cultural contexts. On the one hand, equating climate change solutions with personal behavioral change overlooks structural and systemic factors that account for a large percentage of global emissions. On the other hand, trivializing or ignoring behavioral change downplays the power of individuals to influence social norms and collective action. As we discuss later in the book, integrative solutions to climate change explicitly recognize the connections between individual change, collective change, and systemic change.

Urbanization

A focus on cities and urbanization provides yet another lens on greenhouse gas emissions. By many measures, cities and urban regions account for a disproportionate share of global emissions of greenhouse gases. Just over 50% of the world's population currently lives in urban areas, but cities are responsible for 70–80% of all greenhouse gas emissions (IPCC 2014c). By 2050, roughly 70% of the world's population is expected to live in cities, with the most rapid urban growth occurring in less industrialized countries (United Nations 2018). In addition to demographic factors such as birth rates, urban areas are also increasingly home to migrants from rural areas. People are moving to cities because of both "push factors" and "pull factors." Push factors include declining economic opportunities in rural regions, a process that is fed by globalization and rapid industrialization of agriculture. Pull factors include the possibility of a better-paying job or the promise of a better life in cities. Young people are often attracted to the educational opportunities and cultural offerings of urban areas.

Emissions from cities can be measured in a number of different ways (Dhakal 2010). Territorial emissions produced within a city include those associated with electricity production, heating and cooling of buildings, transportation of goods and people, industrial processes, and other activities that happen within the boundaries of a city or urbanized region. While larger cities would be expected to have higher emissions, all other things being equal, many factors beyond population affect a city's territorial emissions. These include the age and condition of a city's building stock, availability and accessibility of public transportation options, and the types of energy that power a city's electrical grid. In addition to energy and transportation systems within a city, the overall structure of a city's economy also influences its territorial emission profile. As was the case at the national level, cities that are home to large shares of "cleaner" service-sector industries and high-tech knowledge-based industries tend to have lower territorial emissions than those that are home to heavy industrial production (e.g., factories, mills, or refineries). The spatial displacement of emissions is also relevant at the urban level (Marcotullio and McGranahan 2007). Many affluent cities with economies that are highly specialized in knowledge-based sectors are experiencing declining emissions. By contrast, emissions are rising in industrializing cities, particularly in East and South Asia where, in the context of globalization, production and manufacturing processes are increasingly concentrated.

Differences in population density and spatial form also help to explain variation in emissions across cities (Glaeser and Kahn 2010). While rapid urbanization often creates an image of gleaming skyscrapers, bustling shopping districts, and carefully planned suburban communities, there is tremendous variation in urban spatial form across the world. Some cities are more compact and densely populated, while others are sprawling, low-density settlements that are spread over a large area of land. While denser cities are often perceived as "dirtier" and more polluted, the emissions per person in these cities are often much lower than those from cities that are more spread out (Hyde Park Progress 2007). Residents of denser cities live in smaller spaces, meaning lower energy usage for heating and cooling. They are also more likely to have access to urban amenities, including public transport systems. In sprawling cities, which often have limited public transit, there tends to be much greater dependency on mobility via personal automobiles.

More comprehensive methods of accounting for urban emissions, such as calculation of a city's carbon footprint, attempt to take into account emissions that are generated both within and outside of a city to meet the needs and demands of urban residents for fossil fuel energy, food, and consumer products and services of all types (Marcotullio et al. 2014). Cities are key nodes in the transmission of the global ideology of consumption, and the adoption of consumer-oriented lifestyles is growing both within cities and globally (Leichenko and Solecki 2005).

In addition to diets that include more meat, which we discuss in the next section, rising affluence in cities within East Asia, South Asia, and Latin America have led to shifts in modes of transport from bicycles to gas-powered scooters or mopeds to motorcycles and automobiles. This shift has not only contributed to rising emissions of greenhouse gases but is also affecting public health in cities such as Bangkok, Beijing, Kuala Lumpur, New Dehli, and Mexico City, where traffic congestion and air pollution have become ubiquitous. Looking more broadly at sustainability issues, the concept of an **ecological footprint** of cities attempts to capture their total environmental impacts, including pressure on surrounding local and global environments in the form of air and water pollution, land-use changes, and solid-waste disposal (Wackernagel et al. 2006). Such calculations show that the footprints of cities not only extend beyond their administrative borders but are key drivers of emissions and land-use change on a global basis (Seto, Guneralp, and Hutyra 2012).

Urban areas are currently considered to be major "culprits" in climate change and many other environmental problems (e.g., land-use changes, air pollution, biodiversity loss). However, it is also important to recognize the ways in which cities are promoting practical strategies to address rising emissions (Dodman 2009; Bulkeley 2013). Because city governments also have direct control over many factors that influence energy use, they have been centers for many policy innovations that support emissions reductions (Rosenzweig et al. 2018). For example, governments in Paris, Stockholm, Singapore, London, and other cities have implemented **congestion surcharges** which help to both reduce traffic at peak hours and to lower emissions. Other measures that cities may take include changing construction codes to promote greener buildings, mandating that energy comes from renewable sources, and implementing smart city technologies to reduce traffic flows and energy use. For example, Oslo, the capital of Norway, has become a world leader in terms of the use of electric cars by developing infrastructure and incentives that have accelerated the shift from gasoline and hybrid cars to fully electric cars (Aasness and Odeck 2015). Oslo has also committed to a 50% reduction in greenhouse gas emissions by 2020 and 95% by 2030, compared to 1990 levels (Oslo Municipality 2018).

Urban efforts to promote emissions reduction often originate through grassroots efforts led by local residents. The political will to address climate change in cities is often rooted in concerns about environmental justice, especially attention to the unequal burdens of pollution and waste on disadvantaged communities, and the desire for a cleaner local environment. Urban emissions reduction efforts have also been organized through global networks such as ICLEI (International Council for Local Environmental Initiatives) and C40, both of which advocate a role for cities in climate change mitigation. In short, cities and urban areas are becoming important strategic players in global emissions reduction (Castán Broto 2017). Urban environmental justice advocacy and global networks have also influenced the climate justice movement, which we discuss in chapter 10.

Emissions from food systems, agriculture, and land use

Agriculture, forestry, and other land uses represent a major source of greenhouse gases. In fact, they account for more than one quarter of total emissions (IPCC 2014b). Tracing the diverse paths that food products follow from farm to plate helps to illuminate how much food and agricultural systems contribute to climate change. The amount of food traversing the globe has increased exponentially since the 1960s, facilitated by the introduction of refrigerated containers that enable temperature-controlled transport of fruit, vegetables, meat, and dairy products. Reductions in the price for air cargo and the expansion of greenhouses and irrigation schemes have also contributed to these trends. With more than US$1.5 trillion in trade flows, the agricultural sector (including food, animal and vegetable products, oil and fats, and tobacco and beverages) accounts for 10% of the value of international trade (WTO 2015).

Globalization of food systems and consumption habits have meant that diets are becoming increasingly similar, especially as affluence rises. Diets that emphasize meat, poultry, and dairy consumption, as well as increasing consumption of products that rely on global production chains such as coffee (see Box 5.4) and chocolate, play an important role in rising global emissions. Some estimates suggest that meat consumption alone contributes overall to about 10% of greenhouse gases, although the contributions vary depending on production practices (Carlsson-Kanyama and González 2009). Rising demand for meat has also meant that the livestock sector, which currently accounts for about 40% of global agricultural output, is among the fastest growing subsectors of agriculture. At any given time, there are an estimated 1.5 billion cows in the world, as well as 1 billion sheep and 19 billion chickens. Rising meat consumption also contributes to land-use changes when forests are cleared for grazing lands or used to grow animal feed (such as soy), meaning that there are fewer forests to take up or sequester carbon from the atmosphere (McAlpine et al. 2009). Given the outsize role of meat consumption as a driver of emissions, changes in diet are often discussed as a potential strategy for emissions reduction (see Figure 5.6).

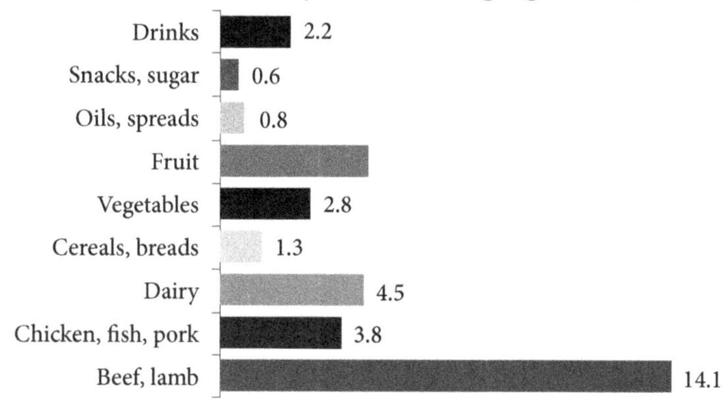

Note: Figures are grams of carbon dioxide equivalents per kilocalorie of food eaten (g CO_2e/kcal). Intensities include emissions for total food supplied to provide each kilocalorie consumed. This accounts for emissions from food eaten as well as consumer waste and supply chain losses. All figures are based on typical food production in the USA. Estimates are emissions from cradle to point of sale, they do not include personal transport, home storage or cooking, or include any land use change emissions

Sources: ERS/USDA, LCA data, IO-LCA data, Weber & Matthews

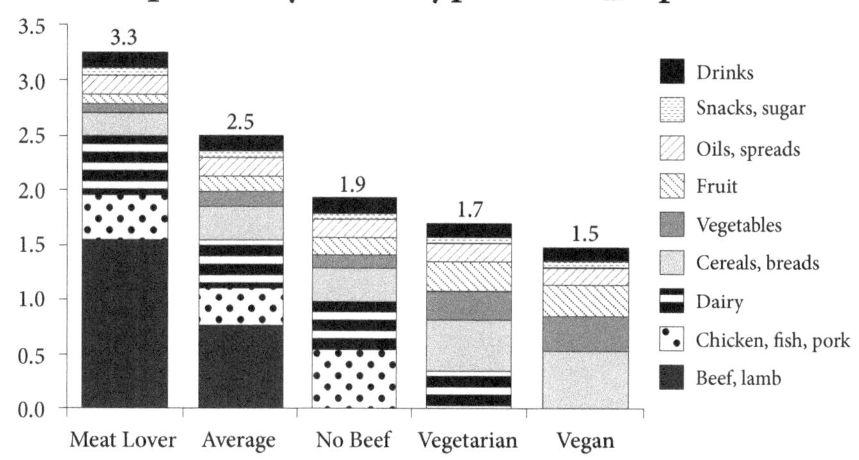

Note: All estimates based on average food production emissions for the US. Footprints include emissions from supply chain losses, consumer waste and consumption. Each of the four example diets is based on 2,600 kcal of food consumed per day, which in the US equates to around 3,900 kcal of supplied food.

Sources: ERS/USDA, various LCA and EIO-LCA data

Figure 5.6 Carbon intensity of eating and footprints by diet type.
Source: http://shrinkthatfootprint.com

Box 5.4 Coffee and climate change

Many people start the day with a cup or two of coffee. In fact, over 2.25 billion cups of coffee are consumed every day. But how and why do our coffee habits contribute to climate change? Coffee beans are grown in subtropical regions of Africa, Latin America, and Asia, and 25 million small producers rely on coffee to make a living. Yet most of the emissions associated with coffee actually come from transportation and consumption. This includes the shipping of coffee beans to processing facilities for roasting, packaging, and distribution to supermarkets all over the world. It also includes the shipping of fresh beans to specialized cafés for small-batch roasting. A full accounting of emissions from coffee consumption would also include fuel used by consumers who drive to grocery stores, coffee shops, restaurants, and other locations where food is purchased, as well as fuel used for cooking or preparing food before it's eaten and the waste produced along the way (Killian et al. 2013). What actions do you think would be most effective for reducing emissions and the carbon footprint of coffee?

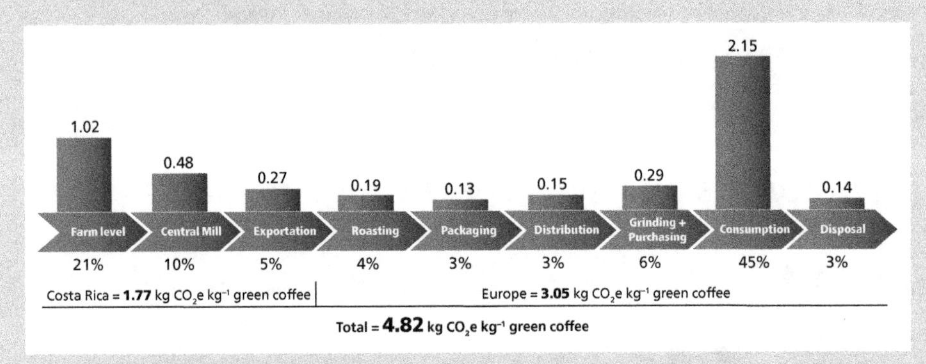

Figure 5.7 Carbon footprint of Costa Rican coffee supply chain.
Source: Killian et al. 2013

Recognition of the role of food transportation in global emissions has helped to popularize other strategies, including increasing reliance on locally grown foods in order to reduce **food miles**. Food miles, which are analogous to carbon footprints, estimate the amount of emissions that are associated with transport of a food item from where it is produced to where it is consumed. According to one online food mile calculator, bananas grown in Costa Rica and sold in the United Kingdom have traveled roughly 8,729 kilometers (5,424 miles). However, prioritizing spatially local food systems does not necessarily ensure that they are socially, economically, and environmentally sustainable (Cleveland, Carruth, and Mazaroli 2015). Affordability of locally produced foods is also a concern, particularly for lower-income groups within industrialized countries. In thinking about changing diets, buying locally grown foods, and other strategies for emissions reductions, it is also important to reflect on social norms and practices that may either support or constrain these efforts, as well as the consequences for equity.

Along with changes in food consumption practices, there are many other strategies for emissions reductions that are related to food. The concept of **drawdown**, popularized by Paul Hawken (2017), is based on the idea that mitigation solutions should focus on maintaining carbon in the biosphere by both reducing emissions and by pulling carbon out of the atmosphere. Many of the strategies for drawdown entail either reducing emissions from food production, for example, by decreasing food waste or increasing carbon sequestration in vegetation and soils. These latter approaches emphasize the role of carbon sinks that can enhance the removal of carbon dioxide from the atmosphere (oceans and freshwater aquatic ecosystems are also carbon sinks). Expanding bamboo cultivation on degraded land, repopulating herds on the Mongolian plains, and eating plant-rich diets are some examples of ways to reduce atmospheric concentrations of carbon dioxide through changes in agricultural and land-use practices (Hawken 2017).

Other land-use approaches target the problem of deforestation. Deforestation is a major concern with respect to the global carbon budget. Tropical and boreal forests sequester large amounts of carbon dioxide, and the intentional removal of these forests for agricultural production or the unintentional burning of these forest compromises this vital function (Seymour and Busch 2016). Preventing deforestation is considered an effective way to avoid emissions, while reforestation and soil management are promoted as ways to sequester carbon. Programs such as **REDD+**, which focus on reducing emissions from deforestation and forest degradation, rely largely on financial incentives and other market-based measures, to motivate farmers and landowners to change their farming practices or keep their forests intact. Such approaches are subject to similar critiques as those raised earlier, including the need for attention to larger social processes that are contributing to rising demand for livestock and other products (Beymer-Farris and Bassett 2012). Concerns about equity in the design and implementation of REDD+ programs have also fostered calls for making mitigation programs more inclusive of, and beneficial for, marginalized individuals and communities (McDermott, Mahanty, and Schreckenberg 2013). For example, there is a need to support alternative livelihood options for farmers. We return to the need for equitable, inclusive, and sustainable responses to climate change in later chapters.

Summary

Atmospheric greenhouse gas concentrations have been rising steadily since the start of the industrial era, and a key challenge facing humanity is how to drastically reduce emissions. Different approaches to both counting and explaining rising emissions point to different ways of addressing this challenge. While compliance with international emissions reduction targets is assigned to nation-states, the distribution and responsibility for emissions is far more complex. As we have shown in this chapter, emissions not only vary among countries but also among firms, cities, households, and individuals. Rising emissions are associated with processes that are not confined to national borders, including globalization, urbanization, and increasing consumption. Industrial firms, households, cities, communities, and other groups play an increasingly visible role in developing and implementing emissions reduction strategies. As we discuss in the next chapter, many of these strategies center on energy use and the energy sector.

Reflection questions

1. Do you think greenhouse gas emissions should be measured by place of production (i.e., as territorial emissions) or by place of consumption?
2. How does your carbon footprint compare to those of your friends and family? Has your own carbon footprint increased or decreased over the past five years? Why?
3. Can you envision an ideology of sufficiency that might replace the ideology of consumption? How might this new ideology influence emissions patterns?

Further reading

Bulkeley, H., 2013. *Cities and Climate Change*. New York: Routledge.

Davis, S. J. and Caldeira, K., 2010. Consumption-based accounting of CO_2 emissions. *Proceedings of the National Academy of Sciences of the United States of America* 107(12): 5687–92.

Hawken, P., 2017. *Drawdown: The Most Comprehensive Plan Ever Proposed to Reverse Global Warming*. New York: Penguin Books.

Rosa, E. A. and Dietz, T., 2012. Human drivers of national greenhouse-gas emissions. *Nature Climate Change* 2(8): 581–6.

Wilhite, H., 2016. *The Political Economy of Low Carbon Transformation: Breaking the Habits of Capitalism*. New York: Routledge.

6 A WORLD OF ENERGY

Let's talk about energy

We live in a world of energy and we talk about it all the time. Whether in relation to having access to an outlet for charging our phones, powering our laptops, or having fuel for our cars or motorbikes, energy is a constant topic of conversation. Yet many of us take our access to energy for granted. We simply expect that we can switch on a light, store food in a refrigerator, listen to digital music, text our friends, and keep up with social media. Unless there's a power outage, we seldom think about the energy systems and infrastructure that fuel our everyday lives. However, because the majority of the energy that most of us use is derived from the burning of fossil fuels, thinking about energy is precisely what we need to do in order to "shed light" on the issue of climate change.

This chapter explores our deep-seated connections to fossil fuel energy and considers potentials for realizing alternative energy futures. We begin by examining the relationship between energy use and greenhouse gas emissions, recognizing that production and consumption of fossil fuels account for the majority of emissions on a global basis. We also discuss energy derived from renewable sources, which is expanding rapidly yet still accounts for only a fraction of total energy usage. The chapter then takes a detailed look at why society is so "locked in" to fossil fuels. Building once again on our discussion of climate change discourses, we explore several explanations for the dominance and persistence of fossil fuels, including those rooted in technology and economics, geopolitics, and political economy. We also draw attention to the vital issues of energy poverty and energy justice, highlighting that many people still lack access to energy and electricity. We conclude by questioning some of the assumptions that both perpetuate fossil fuel usage and limit our thinking about alternatives. We emphasize that there are, in fact, many possibilities for reducing fossil fuel usage and supporting just and sustainable energy futures.

Energy and emissions

Emissions of greenhouse gases began rising on a global basis in the nineteenth century, with the onset of the Industrial Revolution. However, it was not until the end of World War II in 1945 that total energy use began to increase exponentially. At that time, industrialized economies recovering from the war initiated an energy-intensive boom in construction, development, and consumption, as discussed in previous chapters. That energy boom has continued to the present day, with energy-intensive industrialization spreading to countries and regions throughout the world. Most of the energy presently fueling this boom is derived from the burning of fossil fuels, including oil, coal, and natural gas, though there are many other available energy sources such as wind, solar, hydropower, and nuclear energy (see Figure 6.1). If current trends continue, global energy demand is expected to increase by roughly 28% between 2015 and 2040 (United States Energy Information Administration 2017).

Fossil fuel energy sources

The highly concentrated forms of organic compounds in coal, oil, and natural gas are collectively referred to as fossil fuels. Most of the fossil fuels that are consumed today formed during the Carboniferous Period about

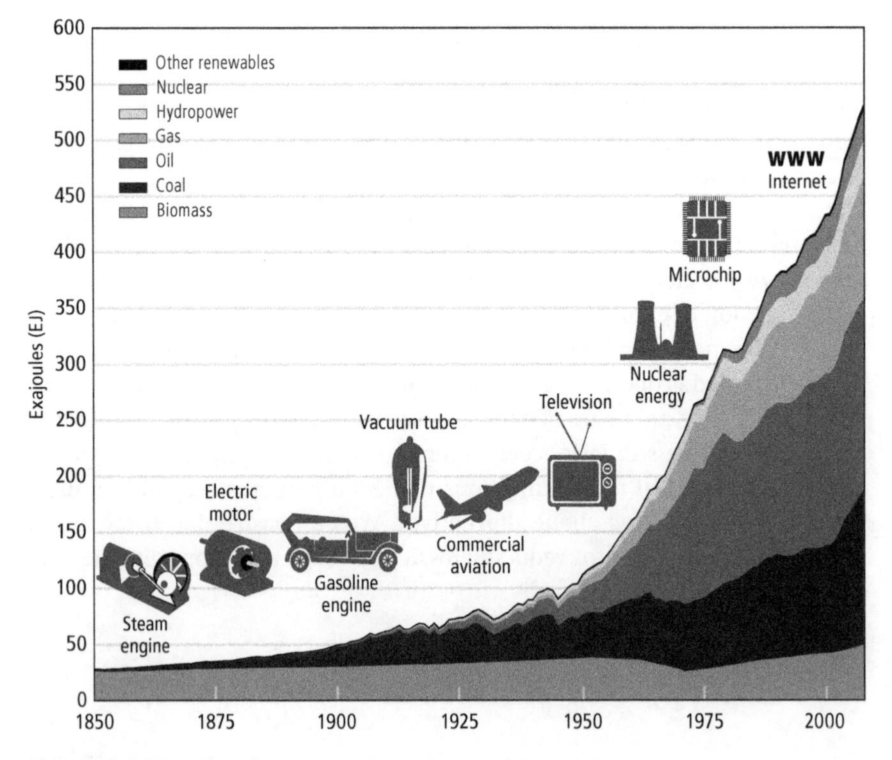

Figure 6.1 Energy sources over time.
Source: GEA 2012

359 to 299 million years ago. Carboniferous means "carbon-bearing," and it describes a period when abundant swamp forest ecosystems sequestered vast amounts of CO_2 from the atmosphere and provided the organic material for much of today's coal and oil. In addition to production of electricity, heating and cooling of buildings, industrial processing, and transportation, fossil fuels are also used in thousands of industrial and consumer products, including petrochemicals, plastics, cosmetics, and clothing. Fossil fuels are considered finite and non-renewable because their formation occurs over millions of years.

Fossil fuels comprise more than three-quarters (78%) of all energy used on a global basis, and they account for roughly 70% of current global greenhouse gas emissions (see Figure 6.2) (REN21 2017). While the combustion or burning of all types of fossil fuels releases CO_2 into the atmosphere, there are some notable differences among the three major fuels – coal, oil and natural gas – in terms of their energy content, emissions profiles, and environmental impacts.

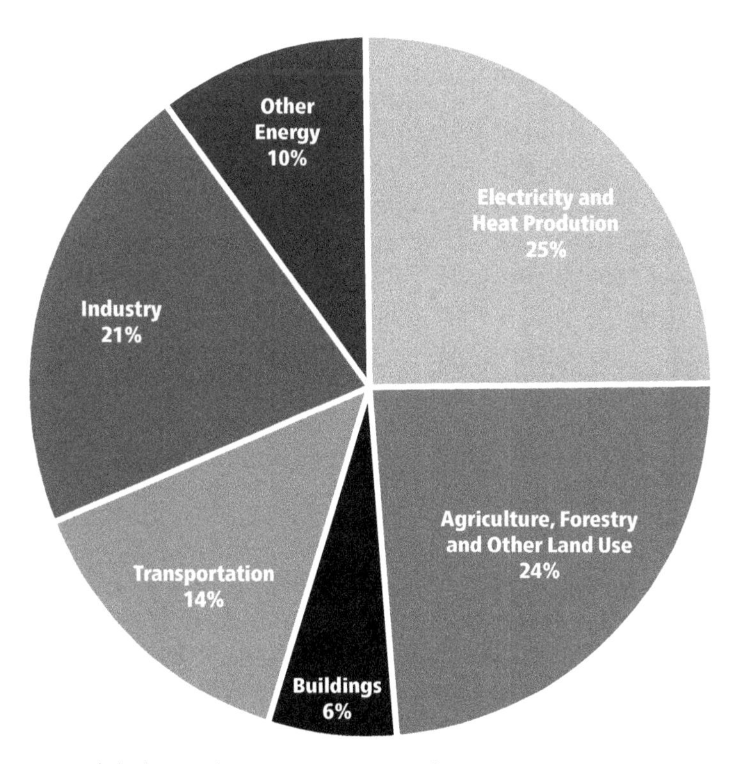

Figure 6.2 Global greenhouse gas emissions by economic sector.
Source: USEPA. Available at https://www.epa.gov/ghgemissions/global-greenhouse-gas-emissions-data

Coal Although coal has been used as a fuel for more than 1,000 years, it was not until the industrial revolution (around 1850) that overall coal consumption began its marked increase, first with the rise of the steam engine for transportation and industrial production, and later to meet the growing demand for electricity (see Figure 6.1). Coal usage has been declining on a global basis in recent years, but it is rising in some parts of the less-industrialized world as a source of fuel for electricity. Coal also continues to play a key role in the production of electricity and heating of buildings in many industrialized countries. In terms of emissions, coal is often described as the "dirtiest" fossil fuel. It not only releases more CO_2 than oil or natural gas per unit of energy produced, but it also releases particulate matter (e.g., soot) and other pollutants, such as sulfur dioxide, which contribute to local air pollution and acid rain.

Oil The rise of oil as a major energy source began in the late nineteenth century with the invention and spread of the internal combustion engine. As automobile use and other forms of motorized transit expanded globally, particularly during the second half of the twentieth century, oil overtook coal as the dominant form of fossil fuel energy. Today, oil is the primary fuel used in the transportation sector, accounting for well over 90% of all fuel used and more than three-quarters of fuel used for road transport alone. There has been considerable debate about if and when society will reach **peak oil** – the time when the maximum level of oil production has been reached, after which production will begin to decline. Some experts think that peak oil will occur within the next several decades, but projections vary depending on production from unconventional deposits, such as oil shale and oil sands (Kuhns and Shaw 2018). In addition to its contribution to greenhouse emissions, oil is also a major source of air and water pollution, and its production and transport increase the risk of damaging oil spills.

Gas The production and use of natural gas have increased remarkably since the middle of the twentieth century. Natural gas, which is primarily methane (CH_4), is used for energy production, heating, and transportation. Natural gas has a relatively higher energy content than other fossil fuels, meaning that less CO_2 is emitted per unit of energy produced. For this reason, natural gas is considered by many to be a "cleaner" fuel in the transition away from oil and coal. However, methane itself is also a greenhouse gas, with a higher global warming potential but shorter atmospheric lifetime than CO_2. Recent technological innovations, including new methods of hydraulic fracturing, or "fracking," have expanded the production of natural gas. The fracking process has come under much scrutiny, however, for its environmental and geologic impacts, including concerns about contamination of water supplies and triggering of earthquakes in seismically sensitive areas (Ellsworth 2013; Vidic et al. 2013).

Alternative energy sources

Prior to the Industrial Revolution, energy for human use was primarily derived from renewable sources, including traditional biofuels, water, and wind. Recognition of the threats associated with climate change and concerns over fossil fuel energy shortages have contributed to growing interest in alternative energy sources as a way to meet rising global demand. Renewable forms of energy, including hydropower, solar power, wind power, geothermal energy, and traditional and modern biomass fuels, currently account for about 19% of global energy consumption. Expansion of renewable energy resources is often seen as key to addressing climate change because these sources provide energy without contributing to net greenhouse gas emissions.

While the production of many forms of renewable energy has been rising in recent years, roughly half of renewable energy still comes from traditional biomass sources. Biomass energy is derived from plants and other living (or recently living) organisms. Biomass energy includes burning of wood, wood products (e.g., charcoal, biochar), and animal dung. It also includes biofuels such as ethanol, which is made from corn or sugar cane, as well as methane fuel derived from municipal or agricultural waste products (methane is released when organic waste decomposes, and it can be captured and used as a source of fuel). Other sources of biofuels, such as soy and palm oil and the jatropha bush, have been heralded as promising energy solutions, yet have not proven to be economically, socially, and environmentally viable (van Eijck et al. 2014).

Other prominent forms of renewable energy, including wind, solar, geothermal, and hydropower, are primarily used for production of electricity. Wind energy, for example, accounts for a substantial share of electricity production in many countries, such as Denmark, Portugal, and Ireland, and has been growing rapidly in China and India. Although solar energy currently accounts for a relatively small share of global electricity production (1% in 2015), it is expanding rapidly in many countries, particularly in China, the European Union, and the United States. Geothermal energy, or heat energy derived from the earth, is also growing in many regions; it accounts for approximately 25% of electricity production in Iceland and El Salvador. In addition to the forms of renewable energy described above, new types of energy are under development and might become available in the future. These include energy from the deep ocean, wave or tidal energy, algae energy, or possibly even energy derived from boron (Hawken 2017).

Hydropower plays a role in supplying renewable energy to millions of people. Hydroelectricity production has doubled over the past three decades, and it is projected to double again by 2050 (Zhang et al. 2018). In Brazil, hydropower generation accounts for more than 76% of the electricity generated, and contributes to approximately 15% of total domestic energy supply. Although hydroelectric potential has been exhausted in many regions of Brazil, plans are underway to construct more hydroelectric dams on the Amazon, which is considered the "new hydroelectric frontier" (Soito and Freitas 2011). However, there are significant constraints on new dam construction, including concerns about the ecological and social consequences of habitat fragmentation along rivers (Anderson et al. 2018). These, together with disrupted river flows and flooding of valleys and canyons for new reservoirs, human and non-human population displacement, competition from drinking and irrigation water needs, and the impacts of climate change on river flows, present significant challenges to sustainable hydropower development (Zhang et al. 2018). As an alternative to large dams, micro hydro projects have been installed in many rural areas. Taking advantage of river flows to generate electricity for households and communities that are not connected to the grid, these micro projects are often a complement to other energy sources (Bracken, Bulkeley, and Maynard 2014).

In thinking about alternatives to fossil fuels, nuclear energy is another important option to consider. Although nuclear energy does not release greenhouse gases, its production requires the mining of uranium, and it is not considered a renewable form of energy. The dangers associated with nuclear energy production, including radioactive pollution and radioactive waste, were widely discussed following accidents at Three Mile Island in 1979, Chernobyl in 1986, and Fukushima in 2011. After these events, many people began to question the benefits of nuclear energy (Siegrist, Sütterlin, and Keller 2014). Germany, for example, decided to curtail its development and use of nuclear power after Fukushima (Renn and Marshall 2016). Overall, nuclear energy has declined in recent years as a fuel source and currently accounts for less than 3% of global energy production (REN21 2017). This share is expected to decline further in coming years as older nuclear plants are decommissioned. Efforts are underway to develop and build safer nuclear plants, yet public sentiment in many countries is against nuclear power (Prati and Zani 2013). Another limitation is the long investment time needed to develop nuclear energy programs and to construct new plants, in contrast to other types of alternative energy resources that can be developed and deployed relatively quickly.

Understanding the synergies, **co-benefits**, and trade-offs among energy sources is vital for developing a sustainable portfolio of energy resources. For example, solar energy provides most of its power during the middle of the day, thus it can be seen as a complementary technology to wind energy, which peaks in the evening or night (Global Wind Energy Council 2017). While alternative energy resources are often discussed as a key option for reducing greenhouse emissions, use of these sources also has co-benefits for human health in the form of reduced particulates and cleaner air. Because these forms of energy are generated onsite using renewable resources such as wind and sunlight, alternative energy can also help reduce vulnerability to disruption of energy transmission due to infrastructure failure. Solar energy offers numerous potential environmental co-benefits, such as utilization of degraded lands and co-location with agriculture (Hernandez et al. 2014). In terms of trade-offs, renewable energy resources can have both social and environmental consequences, particularly for biodiversity (Gasparatos et al. 2017). For example, birds and bats often collide with wind generators, increasing risks to endangered and migratory species. Wind turbines produce noise and influence the landscape, which can lead to resistance from local communities. The production of solar panels involves toxic chemicals such as hydrofluoric acid and sodium hydroxide, uses water and electricity, and creates waste (Tsoutsos, Frantzeskaki, and Gekas 2005). The burning of biofuels produces black carbon, a health hazard that contributes to local air pollution. Yet in relation to the widespread and irreversible impacts of climate change, many such trade-offs can most likely be managed.

Although renewable energy is seen as the next frontier in energy development, fossil fuels are expected to dominate the global energy mix for the next several decades. In the next sections, we discuss some of major explanations for why fossil fuels appear to be so entrenched. These explanations, each of which focuses on a different facet of the logic of fossil fuels, highlight how particular discourses on energy systems and choices may perpetuate carbon-intensive development and potentially limit or undermine consideration of alternative energy options and pathways.

The technical and economic logic of fossil fuels

> The challenge is that nearly the entire energy infrastructure underpinning the modern political economy emits carbon dioxide.
>
> Knox-Hayes 2016: 4

Modern economies are built around the expectation that power will be continuously available to meet demand at all times, day or night. This expectation has co-evolved with technology such that fossil fuels may be understood as part of a comprehensive socio-technical system that is upheld by strong actors and interests. A key argument shared by both the biophysical and dismissive discourses for the continued use of fossil fuels emphasizes their technological and economic advantages over other forms of energy in terms of capacity for storage, portability, and relative costs. Because fossil fuels such as coal, oil, and natural gas store their energy in a dormant form, these fuels can be stockpiled and transported in solid or liquid states over long distances and used to produce electricity, heat, or mechanical energy in different locations. Fossil fuels can also be stored “on site” and burned as needed to meet changing demand levels, particularly during times of peak demand such as extreme heat or cold events. By contrast, wind, wave, and solar energy systems create mechanical energy or electricity at the time that the energy is harvested or captured. Despite ongoing innovations in battery technologies, the lack of ability to store large amounts of energy from renewable sources is a key constraint on their use (Kittner and Kammen 2018).

In addition to greater portability and accessibility of fossil fuel energy in comparison to alternative sources, fossil fuels also presently have production cost and price advantages over most other forms of energy. Energy prices influence which sources are considered viable for producers to exploit and market. The concept of energy return on investment (EROI) permits direct comparison of the costs of energy produced from different sources (Hall, Lambert, and Balogh 2014) (see Figure 6.3). EROI is a ratio of the amount of usable energy produced from a particular energy resource to the amount of energy expended to obtain that energy. Under current technologies, energy returns on investment are generally higher for fossil fuels than for alternative sources. Yet it is important to recognize that EROIs may dramatically change with technological innovation. For example, the cost of producing wind energy and solar energy has fallen precipitously in recent years (IRENA 2018). As EROIs rise for renewable energies, these sources become more economically competitive with fossil fuels.

The technical and economic advantages of fossil fuels are reinforced by a wide range of investments that have been made for more than a century to support fossil fuel development. A sprawling infrastructure, including

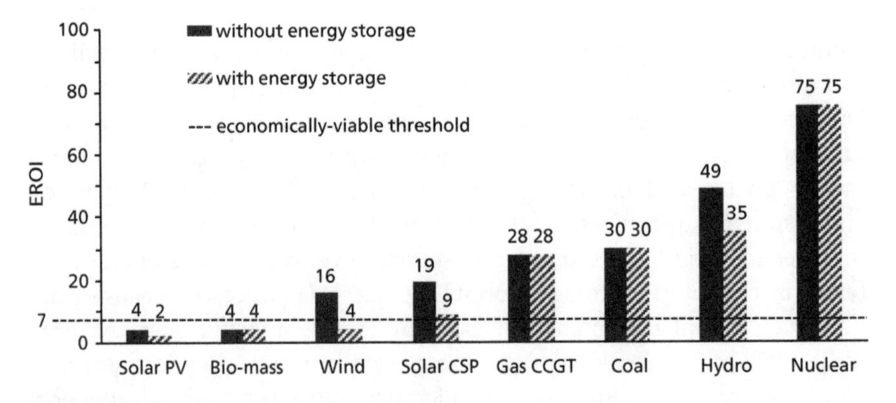

Figure 6.3 Energy return on investment.
Source: Compiled with data from Weißbach et al. 2013 and Conca 2015

tankers, trains, barges, pipelines, and refineries, permits the precise coordination of supply chains, forming the backbone of global oil markets. Although markets for natural gas have historically been regional, new pipelines and technologies that allow liquefaction of natural gas are fostering development of unified global gas markets that are similar to those for oil (Sider and Matthews 2017). These various infrastructure investments also contribute to economies of scale in the production and distribution of fossil fuel energy, lowering costs per unit of energy produced and creating a further barrier to competition from renewables.

The sunk costs associated with investments in particular forms of energy and transport infrastructure systems contribute to a type of **path dependency** known as **carbon lock-in**, where it becomes increasingly difficult and costly for society to change energy systems (Unruh 2000; Seto et al. 2016). The automobile transportation sector illustrates the magnitude, scope, and technical and economic challenges of shifting energy sources. More than one billion automobiles are currently on the road, and automobile use is increasing worldwide. In conjunction with the growing number of automobiles, countries have rapidly expanded road infrastructure and highway systems as well as other infrastructure such as gasoline and diesel fueling stations, parking lots, decks, and garages, all of which support and reinforce the use of automobiles. Although the number of electric vehicles (EVs) is increasing, more than 95% of automobiles currently on the road are powered by gasoline- or diesel-fueled internal combustion engines (Boston Consulting Group 2018). While a large-scale shift to EVs might seem like an obvious solution, efforts to promote their expanded use have been hampered by relatively higher costs, lack of electric fueling stations in some locations, as well as resistance from consumers and auto dealers (Rubens, Noel, and Sovacool 2018). While energy to fuel electric vehicles can, in theory, be derived from renewable sources such as hydropower, wind, or solar, electricity in many locations is currently generated from the burning of coal or natural gas. Unless accompanied by widespread shifts to renewable energy sources, the expansion of electric cars may ultimately contribute to rising energy usage (see Box 6.1).

In addition to direct investment in fossil fuel infrastructure such as roads, railways, and pipelines, carbon lock-in is also associated with institutional structures and government funding of research programs that build particular types of expertise in fossil fuel systems (e.g., petroleum engineering). Government support also takes the form of **fossil fuel subsidies** including favorable regulatory structures and tax laws that promote exploration and mining of oil, gas, and coal. Together, these direct subsidies are estimated to total approximately US$330 billion dollars per year on a global basis (Coady et al. 2017). This provides a powerful economic boost to fossil fuels by lowering the cost of production relative to other forms of energy. All told, the systems,

Box 6.1 Paradox of rising energy efficiency

Energy efficiency is often thought of as a strategy to reduce overall energy demand, and many countries have established energy efficiency requirements for products ranging from automobiles to washing machines. Yet historical examples suggest that energy-saving technological innovations can sometimes *increase* overall energy usage. This phenomenon, termed the Jevons paradox, or "rebound effect," was first identified by economist William Stanley Jevons in the mid-eighteenth century. He observed that technological advancements that increased the efficiency of coal use resulted in higher overall consumption of coal for industrial purposes. The Jevons paradox can also be observed in connection with the invention of the electric light bulb, which replaced candles and kerosene lamps as sources of indoor lighting. While each light bulb uses far less energy than a candle per watt of emitted light, the use of electric indoor lighting expanded so quickly beginning in the early twentieth century that total energy used for electric lighting quickly exceeded the energy savings associated with the switch away from candles (Fouquet and Pearson 2006). Indeed, a hallmark of economic progress in the early and mid-twentieth century has been the electrification and lighting of cities and later of rural and other underserved areas. A current example of the Jevons paradox might be fuel-efficient airplanes that reduce the cost of air travel, thereby increasing the number of passenger-miles flown per year. While evidence of the Jevons paradox is not entirely conclusive, it nonetheless highlights the connections between efficiency and overall energy usage (Sorrell 2009).

Can you think of any products or services that you've consumed or purchased where increases in energy efficiency have increased your overall consumption? Do you think it's possible to increase energy efficiency yet avoid the Jevons paradox?

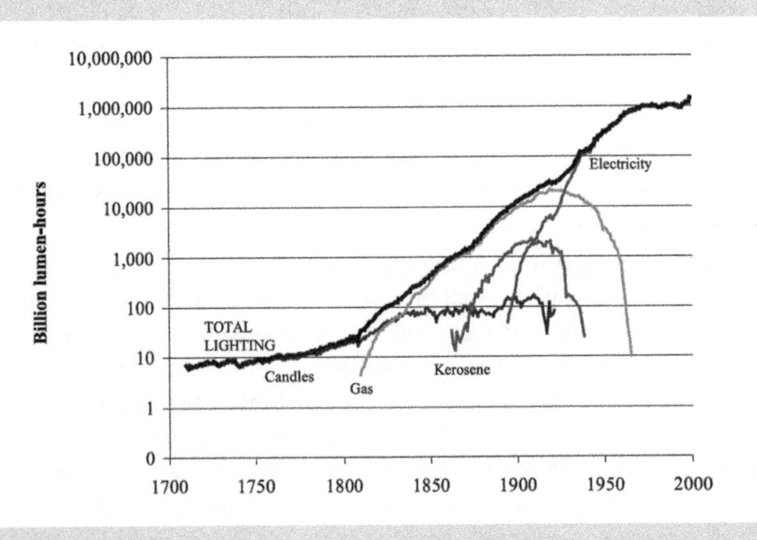

Figure 6.4 Consumption of lighting.
Source: Fouquet and Pearson 2006

investments, and subsidies that support continued reliance on fossil fuels create a situation where producing energy from fossil fuels appears, at least on the surface, to be more cost-effective than producing energy from other sources.

While the technological and economic logic for using fossil fuels may seem somewhat insurmountable, one key caveat to justifications for continued use of fossil fuels is that these arguments do not take into account the associated **externalities**. These externalities include all of the environmental and social costs associated with fossil fuels. Damage to human and ecosystem health as a result of air, water, and soil pollution, damage to public lands and property as a result of oil spills or gas releases, and damage to water supplies from leakage of fracking chemicals into local wells are all examples of externalities associated with fossil fuels. Carbon emissions-related damages, which are not included in the price of fossil fuels, have been quantified through the concept of the **social cost of carbon**. This metric is intended to provide a comprehensive estimate of the costs of damages associated with carbon emissions from fossil fuels, such as reduced agricultural productivity, increased flood damages, greater demand for air conditioning, and so forth (Nordhaus 2017). When all externalities are taken into account, recent estimates suggest that the "true" cost of subsidies for fossil fuels are on the order of US$5 trillion/year, or roughly 6.5% of global GDP (Coady et al. 2017). These estimates are based on a comparison of consumer prices of fossil fuels, supply costs for the fuels, and environmental costs, including impacts on global climate, air pollution, and motor vehicle externalities (Coady et al. 2017). Importantly, the impacts of these externalities are distributed highly unevenly, with some communities and ecosystems more affected than others. Given the essential role of energy costs in economic decision making, pricing fossil fuels to account for their full externalities has the potential to dramatically alter the type and amount of energy that is produced and used.

Geopolitics and national energy security

Fossil fuels have been vital to the geopolitical and economic interests of nations (Mitchell 2011). Geopolitical explanations for the continued exploitation of fossil fuels stress their importance in relation to economic growth and development and national energy security. Countries are increasingly concerned with having a reliable, stable, and secure energy supply that has limited exposure to price volatility or supply disruption (Ang, Choong, and Ng 2015). Energy crises, as experienced during the 1970s following an oil embargo by the Organization of Petroleum Exporting Countries (OPEC) against the United States (in 1973) and decreased oil output (in 1979), have severe economic consequences. However, they also serve as a catalyst for steps to enhance energy security, including diversification of sources, demand-side management, and investments in energy research and development.

Recognition that energy is a geopolitical issue draws attention to how each country's energy interests and desire for energy security can influence its political behavior (see Box 6.2). In many countries, fossil fuel companies have strong ties to national governments, which can further reinforce the fossil fuel status quo. Examples of fossil fuel firms that are (or were) fully or partially state-run industries include Equinor in Norway, Sinopec in China, Lukoil and Gazprom in Russia, Pemex in Mexico, and Petrobras in Brazil. Often the relationship between oil and politics is blurred, with common voices representing both national interests and oil interests. In addition to the major OPEC producers, such as Saudi Arabia, Kuwait, Iran, and Iraq, other major oil producers include the United States, Russia, Brazil, China, Canada, Nigeria, Angola, Venezuela, Mexico, and Norway. Countries that benefit from production and export of fossil fuel energy have a clear interest in the continued sale and use of fossil fuels. Declining global demand for fossil fuels may put economic pressure on oil-producing countries, many of which depend on these revenues for all sorts of public services ranging from infrastructure to education to health care and assistance for the poor. Yet even countries that import most of their fossil fuel energy, such as Japan, South Korea, India, and members of the European Union, still have vested interests in fossil fuels because their built infrastructure, transportation, and industrial systems depend on the production or import of these fuels.

Fossil fuels also represent a source of revenue for less-industrialized countries such as Mozambique, Burma/Myanmar, and Uganda, which have recently discovered new oil and gas reserves and are interested in exploiting them. These new reserves are proposed as a means to offset energy imports, thereby contributing to energy security and providing export revenues that can potentially support economic development. Fossil fuels are often seen as a way to promote development, yet there is no guarantee that the revenues from oil and gas will benefit citizens, rather than enrich a select few. The concept of a **resource curse** applies to situations where exploitation of resources such as oil leads to neither economic benefits nor positive development outcomes (Watts 2018). In many cases, partnerships with large multinational fossil fuel companies have been established to explore, extract, and develop fossil fuel resources in less-industrialized countries. For example, US-based Exxon-Mobil, Italy-based UNI, and British-based BP are among the key investors in a newly developed project to export natural gas from Mozambique. As we mentioned in chapter 5, the idea that countries should "give up" these opportunities and leave fossil fuels in the ground raises questions of equity and fairness, given income disparities and the long history of fossil fuel exploitation in industrialized countries.

Box 6.2 The geopolitics of energy in the Arctic

The Arctic is rapidly becoming a new frontier for energy exploration. As Arctic sea ice melts, reserves of oil and gas in the Arctic Ocean are becoming more accessible and easier to transport via newly open shipping routes (Leichenko and O'Brien 2008). McGlade and Ekins (2015) estimate that, as of 2010, there were 100 billion barrels of oil and 35 trillion cubic meters of gas within the Arctic Circle that have yet to be exploited. Many countries and corporations are eyeing these new opportunities for exploiting oil and gas resources in the Arctic as a potential economic bonanza, and some have increased their military presence and activities in the area. Growing interest in the mining of oil, gas, and minerals has the potential to contribute to future geopolitical conflicts, especially as Arctic countries such as Canada, Norway, Russia, and the United States vie for control of newly accessible resources (Holder et al. 2015).

Some analyses suggest that the idea of a "new Cold War" in the Arctic is overblown, particularly when the actual stakes, interests, and

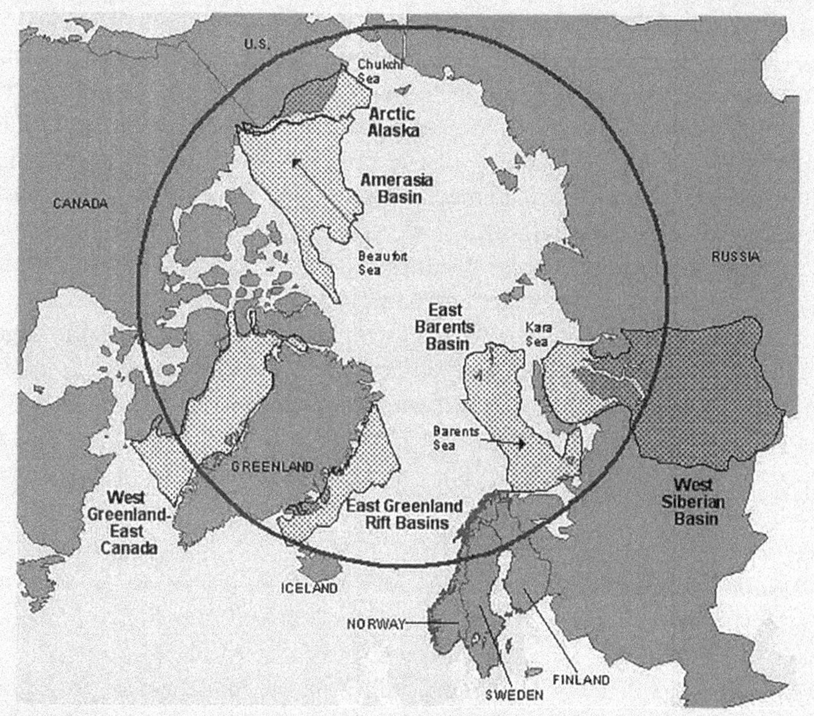

Figure 6.5 Arctic oil and natural gas resources.
Source: US EIA 2012, based on USGS

institutional capacities of those involved are taken into consideration (Keil 2013). The Arctic Council, an intergovernmental forum with eight member states, six indigenous permanent participant organizations, and a number of observer states and organizations is an important venue for reducing the potential for conflicts. The Council's mission emphasizes promotion of cooperation, coordination, and interaction on Arctic issues, especially related to environmental protection and sustainable development.

Do you think that fossil fuels that become more accessible as a result of climate change should be exploited? If so, who do you think should control access to these resources? If not, how do you think these areas can be protected from fossil fuel mining?

While prior investments and dependencies on fossil fuels can exert powerful pressure to maintain the status quo, global patterns of energy production and consumption are changing rapidly. In addition to increasing competition from renewable energies, recognition of the health and environmental costs of fossil fuels is also changing the geopolitical calculations for some countries. China, for example, which has traditionally relied on coal for its energy production, faces severe air pollution problems in many of its cities. This has led to policies and action plans that recognize the need to transform their current energy systems (Sheehan et al. 2014). With substantial potential to develop solar, wind, and other renewable resources, China is expected to be able to meet much of its future energy demand with renewable energy (Liu et al. 2011). However, in a globalizing world, the spatial displacement of emissions also needs to be taken into account. While China has been reducing emissions at home by closing coal-fired power plants and developing renewable energy resources, it has also been investing in high-emitting activities in Latin American countries. To secure long-term access to energy, Chinese public oil companies have been buying or investing in oil and gas reserves in Brazil, Peru, Ecuador, and Venezuela (Edwards and Roberts 2014). Such examples illustrate that the geopolitics of energy are complex and dynamic and that future trajectories of energy production are highly intertwined with globalization.

While perpetuating the fossil fuel economy may ensure energy security in the short run, it also contributes to economic, social, and environmental risks in the long run. Moreover, many countries that have oil, gas, and coal resources to exploit are also highly vulnerable to the impacts of climate change. The Caribbean country of Trinidad and Tobago, for example, is an oil producer, yet it is also among the island nations most exposed to the impacts of sea-level rise. As such, it is vulnerable to both the loss of fossil fuel income from climate change mitigation actions, and to the direct impacts of climate change. This situation presents difficult contradictory choices for leaders who may be held accountable for decisions, one way or another (Hughes 2017).

The political economy of fossil fuels

Political economy explanations for the continued dominance of fossil fuels draw attention to the ways that energy systems have been shaped by the logic of capitalism and its imperative for continued economic growth. Rooted in the critical discourse, these approaches emphasize the power and influence that the fossil fuel industry wields in promoting continued production, distribution, and use of fossil fuels and draw attention to issues of justice and equity. This discourse reminds us that there is nothing natural or inevitable about the dominance of fossil fuels and points to the political and economic barriers that have prevented the expansion of renewable energy resources.

The fossil fuel industry and vested interests

Recognition of the size, scope, and political influence of the fossil fuel industry provides an important perspective on the question of why fossil fuels are so entrenched. By any measure, fossil fuel companies are major players in the global economy. Listings of top firms by revenue consistently place oil and gas companies among the largest firms in the world. In 2017, for example, four of the ten largest firms in the world were oil and gas companies (see Table 6.1). Of the remaining six companies in the top ten, three are tightly linked to oil and gas, including two that manufacture automobiles and one that produces electricity (Fortune 2018). Large fossil fuel companies and their investors can wield power via contributions to political campaigns and the lobbying of government officials to promote continuation of direct financial support for fossil fuels, such as tax breaks for new exploration, or to block efforts that support the research and development of alternative energy sources.

Fossil fuel firms are not only highly profitable for shareholders and investors, but they also employ millions of people worldwide. Estimates of total global employment in fossil fuels vary widely, but more precise employment counts are available for individual countries. Within the United States, for example, direct employment in fossil fuel industries is estimated to be approximately 1.1 million. Many millions more are employed in firms that support these industries, including transportation, production of mining equipment, logistical support, offshore technologies, and so on (United States Department of Energy 2017). In Norway, an estimated 185,300 people

Table 6.1 Ten largest firms globally in 2017

The Top 10		Revenues (US$M)
1	Walmart	$485,873
2	State Grid	$315,199
3	Sinopec Group	$267,518
4	China National Petroleum	$262,573
5	Toyota Motor	$254,694
6	Volkswagen	$240,264
7	Royal Dutch Shell	$240,033
8	Berkshire Hathaway	$223,604
9	Apple	$215,639
10	Exxon Mobil	$205,004

Source: http://fortune.com/global500

were employed either directly in the petroleum industry or indirectly in petroleum-related industries in 2016. Although this still represents approximately 7% of total employment in the country, it denotes a marked decrease from the 232,000 employees in 2013, a decline that resulted from falling oil prices (Norwegian Petroleum 2018).

The incentive for fossil fuel producers to continue production is reinforced by their ownership and control of fossil fuel reserves. These entail rights to mine assets that are still in the earth, including under the ocean. Proven fossil fuel reserves, which encompass those that could be mined with current technologies, are valued at more than US$25 trillion. While market conditions and technological factors influence decisions on whether or not to exploit these resources, estimates based on a carbon budget that is consistent with limiting global warming to 2°C suggest that most of these reserves will ultimately need to remain in the ground in order to avoid catastrophic climate change (McGlade and Ekins 2015) (see Box 6.3). These reserves thus have the potential to become **stranded assets** for fossil fuel companies, resulting in large economic losses. In addition to the risk of stranded assets, there is also the possibility of successful litigation against fossil fuel companies for damages from climate change (Byers, Franks, and Gage 2017). While lawsuits against fossil fuel companies by countries, states, cities, and other entities have thus far proven unsuccessful, the possibility of successful litigation or changes in legislation around climate change liability is a serious financial risk for fossil fuel companies.

Drawing attention to how fossil fuel dependency is maintained by vested interests may appear as simply an effort to portray fossil fuel firms as the "bad guys." Indeed, as we discussed in chapter 3, the film and book *Merchants*

Box 6.3 Do the math!

> When we think about global warming at all, the arguments tend to be ideological, theological and economic. But to grasp the seriousness of our predicament, you just need to do a little math.
>
> Bill McKibben, 2012

It all comes down to numbers. At least that's what climate activist and author Bill McKibben argued in 2012. To understand "our precarious – our almost-but-not-quite-finally hopeless – position," McKibben suggests that we need to look at three simple numbers: 2°C, 565 Gt, and 2,795 Gt (McKibben 2012). Here is how they relate: The 2°C temperature goal will be achieved by limiting the amount of additional carbon dioxide emissions to only 565 Gt. However, the known reserves of coal, oil, and gas that fossil fuel companies controlled as of 2012 would, if burned, release 2,789 Gt of carbon dioxide emissions. In addition to drawing attention to the "simple" math behind fossil fuel consumption, the "Do the Math" campaign by 350.org also advocated divestment from fossil fuel companies.

Fossil fuel **divestment campaigns**, which advocate that investors withhold their capital from fossil fuel companies, have led to decisions by many cities, universities, and pension funds to pull out of or discontinue investing in coal, oil, and natural gas. In addition to a goal of making continued production unprofitable, the campaigns may also have the effect of changing expectations and social norms for market investments (Ansar Caldecott, and Tilbury 2013).

Do you know if there is a fossil fuel divestment campaign at your school or university or in your city? Do you think that fossil fuel divestment campaigns are likely to be effective in reducing reliance on fossil fuels? Why or why not?

of Doubt present extensive evidence that oil companies have actively misled the public by challenging the science of climate change (Oreskes and Conway 2010). However, it is also important to remember that maintenance of high-energy lifestyles is deeply intertwined with the culture of consumption. Consumers who benefit from cheap energy also have vested interests in maintaining fossil fuels, as it allows travel, mobility, and all of the "stuff" that contributes to high-energy lifestyles, as discussed in chapter 5.

Energy poverty and energy justice

Political economy approaches point out that the current energy regime does not work for everyone. The notion of energy justice recognizes the uneven distribution of both the benefits and costs of energy production and consumption (Jenkins et al. 2016). At a fundamental level, energy justice is about having enough energy to meet basic needs, such as cooking, heating or cooling, transport, social services, and support for livelihoods. Beyond basic needs, energy is also vital for development. As digital communication technologies such as cellphones and computers spread, energy access and availability become increasingly important for information, communication, and entertainment. A nation's development status is closely related to the extent and quality of its access to energy, including how power supplies and energy resources are distributed and used. The importance of energy in human development and well-being has been widely recognized and is represented in Sustainable Development Goal 7, which aims to ensure access to affordable, reliable, sustainable, and modern energy for all.

As much as fossil fuel energy has contributed to development, this development has been highly uneven. Many households and communities have yet to see the benefits of development, whether in relation to basic human needs, such as food, water, and shelter, or energy access. According to the International Energy Agency (IEA) (2017), 1.2 billion people worldwide (17% of the global population) still lack access to electricity, despite modest improvements in recent decades. Households in the poorest and least economically developed countries may experience **energy poverty**, which includes a lack of access to modern energy services, such as electricity, fuels, and stoves that do not cause indoor air pollution (González-Eguino 2015). The burdens of energy poverty often fall disproportionately on poor women (Denton 2002). In many rural regions of Africa and South Asia, poor women spend considerable time acquiring or gathering wood, dung, or other fuels to meet their family's energy needs. For poor women and their families, a lack of electricity for lighting makes it difficult to work, study, or care for children in the evenings or at night, and indoor cooking creates pollution and health problems. Women without access to electricity also have little interaction with media and communication technologies.

Energy poverty is not limited to poor countries, and many people within energy-rich, industrialized countries also face challenges in meeting their energy needs. For example, in many parts of the United States, energy bills for residential needs account for a significant share of income for some of the poorest residents (Teller-Elsberg et al. 2016). Paying for heating during cold weather and for the additional electricity used for air cooling during a heat wave is often a major financial burden for low-income individuals and households.

A focus on energy justice brings the issue of winners and losers to the forefront of energy policy discussions. Energy justice is related to the concept of environmental justice, which has raised awareness of the impacts of siting of power plants and other industrial facilities in low-income communities and the disproportionate exposure of such communities to air and water pollutants from these facilities (Schlosberg 2004). Concerns about energy justice arise when efforts to address climate change are seen as having a disproportionate effect on the poor. Implementation and siting of alternative energy infrastructure, such as wind and solar farms, is also sometimes viewed through a lens of energy and environmental justice. The siting of wind farms in the United Kingdom, for example, has met strong resistance from some local communities. The environmental justice concerns in these cases have to do with lack of voice or local input in the decision-making processes, including perceptions that local objections to siting wind turbines in culturally valued landscapes have been ignored or delegitimized by outside "experts" (Jenkins et al. 2016). A growing movement for climate justice, which we discuss in chapter 10, draws attention to issues of equity both in vulnerability to climate change and in climate change policies and responses.

Energy solutions and energy futures

Reducing fossil fuel usage is a key element of climate change mitigation and many strategies are already underway to accomplish this goal. Largely rooted in the biophysical discourse, these approaches center on regulating, pricing, or taxing fossil fuels. Regulations, taxes, and trading schemes share the common goals of reducing the demand for fossil fuel energy or replacing these sources with renewable energy through market-based interventions (Aldy and Stavins 2012). Following from the economic logic of fossil fuels discussed above, raising the costs of fossil fuels is expected to provide financial incentives for energy users to change behavior, whether by reducing fossil fuel consumption, increasing efficiency, or switching to alternative energy sources. While a comprehensive review of proposed energy solutions is beyond the scope of this book, it is nonetheless important to be aware of some of the key strategies that are currently on the table.

One of the most widely discussed emissions reduction strategies entails taxes on fossil fuels. Arguments in favor of what is sometimes referred to as a **carbon tax** are partly justified by the idea that such a tax would take into account a wide variety of externalities associated with these fuels. Such a tax would increase the price of fossil fuels to better reflect their true costs. A carbon tax would also make alternative forms of energy more economically competitive with fossil fuels and potentially encourage a shift towards renewables. Higher fossil fuel energy prices would also serve to promote energy-saving socio-technical innovations and practices such as cycling or ride sharing. While carbon taxes may seem like a simple answer, policy interventions that raise fuel prices are often fiercely resisted because they will affect profitability of the fossil fuel industry, directly increase energy costs for businesses and consumers, and potentially slow rates of economic growth. Carbon taxes can also be regressive, hitting hardest those who are least able to pay. Another concern with respect to carbon taxes is what should be done with the revenue generated by carbon taxes. Should these revenues be used to subsidize public transportation, refunded to consumers, redistributed to low-income households, or invested in renewable energy programs? Yet even when there is political support for carbon taxes, they can be challenging to implement effectively. Norway, for example, has implemented a relatively high carbon tax on gasoline since 1991, but research shows that it has had a relatively small effect because of tax exemptions (especially for energy-intensive industries) and **inelastic demand** in the sectors where the tax has been implemented (Bruvoll and Larsen 2004).

Another prominent approach to reducing emissions associated with fossil fuels entails regulation of energy producers. While the aim of such regulations is to reduce carbon dioxide and other greenhouse gas emissions, the proposals are often based on examples of successful historical efforts to reduce transboundary air pollution. For example, regulation of sulfur dioxide emissions by coal-burning factories and energy plants was used to successfully address the problem of acid rain during the 1980s. The regulations that were put into place limited the burning of high-sulfur coal, which was damaging lakes, streams, and forests and affecting fish and other wildlife. Similar regulations may be used to limit allowable emissions from power plants or to mandate that a particular share of energy production comes from alternative sources, such as wind and solar energy. Regulations may be enforced by directly targeting sectors that are responsible for large shares of emissions, such as transportation. China, for example, has recently put regulatory limits in place for gasoline-powered automobiles (e.g., limited availability of licenses, high license fees) with the explicit aim of shifting a large share of its automobile fleet to electric-powered cars (*Bloomberg News* 2017).

The creation of **carbon markets** for the purchasing, selling, and trading of emissions credits is an alternative to regulations that impose strict limits on emissions for individual energy producers. These market-based approaches, which are used by the European Union as well as a number of US states, establish a limit or cap on total allowable carbon emissions for energy producers and industrial plants. Specific emissions allowances or permits are then allocated to each participant in the market. Participants who exceed their emissions allowance must purchase additional allowances from those producers who are not using their full allocation. Emissions trading schemes provide an economic incentive for high emitters to reduce emissions in order to avoid the costs associated with purchasing permits. Emissions trading is sometimes argued to be more economically efficient than strict regulation of emissions quantities. Producers who would experience very high costs associated with reducing emissions are able to purchase emissions permits from producers who can reduce their emissions at lower costs (Perdan and Azapagic 2011).

In addition to regulatory mandates, many countries and states have begun subsidizing production of alternative energy sources through direct payments (grants) or tax breaks for producers. The goal of these subsidies is to reduce the relative cost of alternative energies in comparison to fossil fuels in order to make these products more attractive to all types of consumers, including low-income households. Nonetheless, there are likely to be both winners and losers from energy transitions. Within Germany, for example, the implementation of the *Energiewende*, or transformation to renewable energy, has contributed to a rapid increase in decentralized solar energy production by households, where the energy produced feeds back into the grid. Yet, as with carbon taxes, the effects of such policies can be uneven. While the *Energiewende* has benefited consumers who produce energy to sell (referred to as **prosumers**), other households have experienced the negative impacts of rising electricity prices (Fischer et al. 2016).

Market-based climate policies have had limited effectiveness to date in reducing the growth of emissions (van Renssen 2018). In addition to aforementioned disagreements over emissions counting, the lack of political will, difficulties with enforcement, and many other factors have constrained the effectiveness of these efforts (Perdan and Azapagic 2011). While identifying options to improve these efforts holds some promise, those working outside the biophysical discourse point to many other problems with these approaches. The critical discourse, for example, questions whether strategies rooted in the market system will simply replace fossil fuels with other energy sources without challenging high levels of consumption or resource degradation that are inherent in capitalist growth models (Wilhite 2016). The dismissive discourse sees these approaches as adding unnecessary regulations and costs that will constrain innovation and limit economic growth (Stern et al. 2016).

Objections from both the critical and dismissive discourse share a concern about the potentially uneven and disruptive consequences of large-scale shifts away from fossil fuel production. In addition to concerns about effects on low-income consumers, other likely "losers" include oil-producing nations, fossil fuel firms, and workers in these industries. As noted earlier, the fossil fuel industry and related businesses account for millions of jobs globally. For those who hold jobs related to coal, oil, or natural gas, a shift from fossil fuels to renewable energy directly threatens employment and livelihoods. While jobs related to solar and wind energy, energy efficiency, smart-grid technology, and battery storage are growing more rapidly than jobs in fossil fuels, renewable energy jobs are often found in different geographic locations and require different skill sets than jobs in fossil fuels. Moreover, jobs are often tied to people's culture, history, and identities, and adapting to a new form of livelihood or occupation can be challenging, particularly after many years of working in one sector.

Given the impacts and risks associated with climate change, which we explore in the next chapters, it is worth questioning how "locked in" fossil fuels really are. The integrative discourse reminds us that underlying our reliance on fossil fuels are the beliefs, values, and mindsets of a myriad of actors within both the private and the public sectors. This includes our own mindsets. As such, perhaps our greatest energy challenge stems not from fossil fuel dependency but from our inability to imagine alternatives (Milkoreit 2017b). While it is, indeed, difficult to imagine a world where development is not tied to the political economy of fossil fuels, there is little doubt that the energy landscape will change over the next century. What is now taken for granted about energy may appear as strange or incredible to future generations. They may wonder: why did people burn fossil fuels when valuable hydrocarbons could have been used in more meaningful ways? Did they not see the potential for algae power, magma power, or even human power? Were people really so addicted to fossil fuels – and to the idea of economic growth – that they could not even imagine replacing the carbon economy with more sustainable alternatives? We will return to the importance of "seeing things differently" in chapter 10, when we discuss integrative solutions that both address climate change and support transformations to sustainability.

Summary

The expectations that energy will be affordable and supplies will continually increase to meet ever-growing needs are foundational to modern industrial societies. Yet recognition that fossil fuel energy extraction, production, and consumption are key drivers of climate change has led many to question these assumptions, as well as the long-term viability of this model of development. In this chapter, we have discussed a variety of explanations as to why fossil fuels are so dominant, and why replacing them with other energy resources is no simple matter. We have demonstrated that the energy status quo is embedded in a social-technical system rooted in a particular logic and reasoning, supported by vested interests, and maintained by geopolitics. Yet we have also emphasized that an energy transition is underway, and that numerous alternative pathways exist. A lingering question, however, is whether and how societies can rapidly transform energy systems while also providing sufficient energy to ensure well-being for all.

Reflection questions

1 Where does your energy come from? Do you know what percentage of it is based on renewable energy resources? Are you aware of any policies or plans in your community, state, or country to increase use of renewables?
2 Which strategy or strategies do you think might be most effective for reducing greenhouse gas emissions from fossil fuels?
3 Do you think that reducing emissions requires challenging the dominance of fossil fuels? If so, what actions do you think might be effective for accomplishing this?

Further reading

Aldy, J. E. and Stavins, R. N., 2012. The promise and problems of pricing carbon: Theory and experience. *The Journal of Environment & Development* 21(2): 152–80.

Coady, D., Parry, I., Sears, L., and Shang, B., 2017. How large are global fossil fuel subsidies? *World Development* 91: 11–27.

Hughes, D. M., 2017. *Energy without Conscience: Oil, Climate Change, and Complicity.* Durham, NC: Duke University Press.

Jenkins, K., McCauley, D., Heffron, R., Stephan, H., and Rehner, R., 2016. Energy justice: A conceptual review. *Energy Research & Social Science* 11: 174–82.

Knox-Hayes, J. K., 2016. *The Cultures of Markets: The Political Economy of Climate Governance.* New York: Oxford University Press.

7 CLIMATE CHANGE IMPACTS

Experiencing climate change

Imagine waking up one morning to discover that most of your town is under water. This is precisely what happened to residents of Manchester, United Kingdom, on December 26, 2015. On that day, Manchester and the surrounding area received the equivalent of one month of rainfall. In December 2015, many rivers across northern England exceeded their historical record level by more than one meter, contributing to heavy flooding throughout the region. In Manchester, shops and roads were under water, power was out in most areas, cash terminals were not working, and thousands of homes were damaged. Although the UK flooding cost an estimated US$1.9 billion, no lives were lost (United Kingdom Environment Agency 2018).

The impacts of climate change are already being experienced worldwide. Unprecedented storms and record-breaking weather events seem to occur on a regular basis, and phrases such as "superstorm," "monster cyclone," "killer heat wave," and "deadly wildfire" have become part of our everyday vocabulary. Yet the impacts of climate extremes are experienced differently within and across regions, communities, and sectors, depending on the social context. In the month prior to the Manchester floods, torrential monsoon rains associated with one of the strongest El Niño years on record caused massive flooding in the Indian city of Chennai. Roads were submerged and subways were waterlogged, transport within the city was nearly impossible, and more than seventy people lost their lives (BBC 2015). In total, flooding in southern India during that month displaced over 1.8 million people and contributed to more than 500 deaths. The amount of damage from the Indian flooding was estimated at approximately US$3 billion (Narasimhan 2015).

In this chapter, we explore the past, present, and potential future impacts of climate change. We begin by considering the direct impacts of climate change, including its gradual effects, such as the changing length of growing seasons, decreasing snow cover, and rising temperatures and sea levels, as well as changes in the magnitude and/or frequency of extreme events. We also examine some widely used methods for estimating and projecting the impacts and exposures on different sectors and regions. We then consider the less tangible impacts including consequences for cultural values, cultural heritage, and traditional ecological knowledge. By highlighting impacts, exposures, and risks, we draw attention to what is at stake with climate change.

Assessing the impacts

Climate change is transforming human and natural systems throughout the world. While there are many ways to study the impacts of climate change, every approach raises some common questions: How are impacts defined and measured? Where are impacts likely to occur? How do we know that the impacts we are seeing are actually related to climate change? How can we project future climate impacts? These are important questions, and their answers have direct implications for policies, planning, investments, and other actions taken today and in the coming years. **Climate impact assessments** – an umbrella term for a wide range of impact studies – provide guidance and tools for decision makers, resource managers, communities, and others to plan for climate change. They also provide support for policies that reduce climate change risks and vulnerabilities, including mitigation of greenhouse gas emissions.

Impact assessments

Impact assessments are intended to improve understanding of how systems and sectors will be directly or indirectly affected by changing climatic conditions. Generally associated with the biophysical discourse, they have been applied to areas such as health, agriculture, forestry, water resources, coastal zones, species, and biodiversity (see Figure 7.1). Most impact assessments focus on outcomes that can be observed, measured, and projected (Füssel and Klein 2006). They typically begin by identifying which facets of climate change may affect a particular sector, region, or ecosystem. Next they trace the consequences and societal implications under different climate change scenarios. For example, changes in the amount and timing of annual rainfall can influence household and municipal water supplies and water quality, which could create competition for water between agricultural and municipal interests. In the case of coastal zones, warming oceans and rising sea levels can damage coral reefs, destroy marsh habitats and breeding grounds for some species of fish and migrating birds, and contribute to erosion as well as flooding of properties and infrastructure. For coral reefs, the impacts of warming waters are compounded by ocean acidification, which weakens the structure of corals and has negative effects on reef ecosystems, with further implications for fisheries (Anthony 2016).

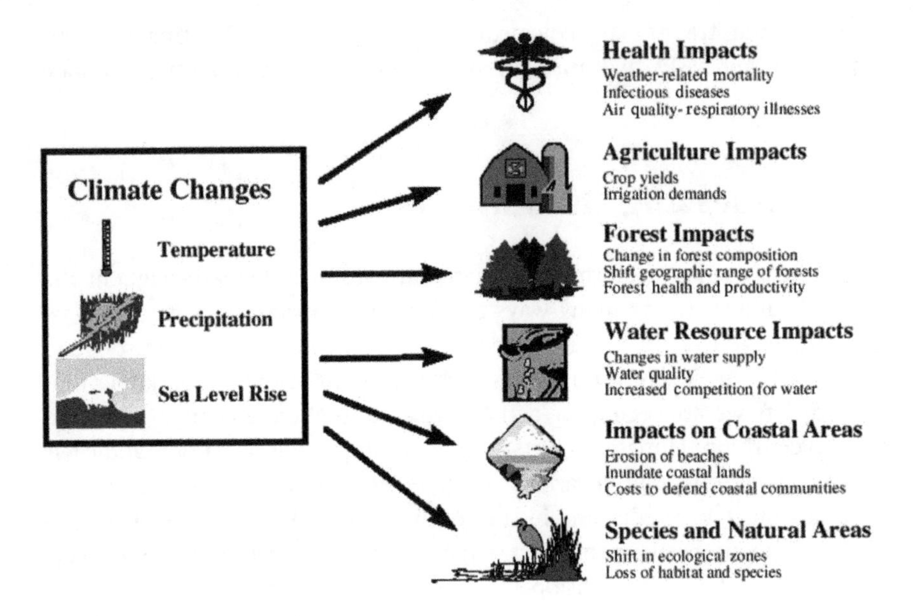

Figure 7.1 Potential climate change impacts.
Source: US EPA

Impact studies often employ different types of mathematical models to conduct "what if" experiments based on diverse scenarios of climate change. In crop yield models, hydrological models, glacier balance models, and other such models, one or several variables are changed to identify the sensitivity of the system to climate change and to investigate outcomes for specific sectors. For example, agricultural researchers studying the impacts of climate change on crop production can explore how evapotranspiration rates might change with warmer temperatures, which can then be used to assess the need for irrigation (Saadi et al. 2015) (see Box 7.1).

Scenario-based studies can also explore how climate changes associated with different emissions trajectories might influence biophysical phenomena, such as forest growth or availability of water for hydropower production. For sectors such as tourism and recreation, climate impacts can be assessed by modeling how consumer demand for particular activities may be influenced by warmer temperatures or loss of snowpack. In the case of ski tourism, warmer temperatures are projected to result in dramatic declines in ski opportunities and revenue in many areas (Steiger et al. 2017). Other sectors, such as health care and apparel, may see both positive and negative changes in future patterns of consumer demand in response to **climate shocks and stresses**, including changing needs for medicine and health products, and shifting preferences for clothing and other goods and services (Leichenko 2018).

Box 7.1 Climate change impacts on agriculture

Have you ever considered what climate change means for the food you eat? For your favorite fruits and vegetables, bread, and even chocolate? As we discussed in earlier chapters, agriculture and food production contribute significantly to the rising emissions that are driving climate change. The agriculture sector is not merely a driver of change but is also on the front line in terms of climate change impacts. Harvests and yields of all types of cereals, fruits, vegetables, and other crops are sensitive to climatic conditions, particularly variations and extremes in temperature and rainfall.

Climate impact assessments are often used to project the direct effects of climate change on crop yields, including how warmer temperatures, higher atmospheric CO_2 concentrations, and changes in the amount and timing of rainfall will influence production of major food crops such as wheat, maize, rice, and soybeans (Zhao et al. 2017). In these assessments, crop production models can also be integrated with global trade models to explore how decreases in productivity and output in one region of the world as a result of climate change can be offset by importing surpluses from another region. These integrated models can also explore what happens to global crop outputs if decreases in productivity occur in several regions simultaneously.

The impacts of climate change for agriculture are likely to be especially severe in semi-arid regions, where crop production and livestock depend heavily on

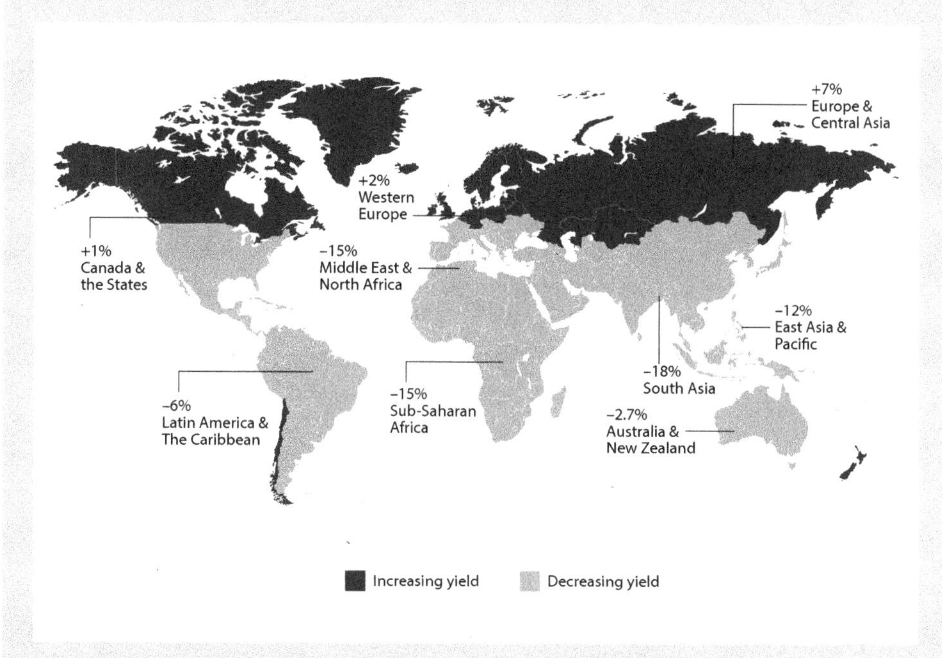

Figure 7.2 Climate change and projected agricultural yields.
Source: World Bank Group, 2010

the amount and timing of rainfall. In particular, rain-fed agriculture in the semi-arid tropical regions of Africa, Asia, and South America is expected to experience large decreases in crop yields (World Bank 2010). Although crop production in some high-latitude regions (for example, Canada, Russia, and Scandinavia) is expected to initially benefit from climate change, higher temperatures and greater rainfall variability may also contribute to the spread of new pests and crop diseases. Climate change may also reduce the quality of food crops, including fruits, vegetables, herbs, and tree nuts (Ahmed and Stepp 2016).

While crop models provide clues as to the direct impacts of climate change on agricultural production, changes in crop yields are just one dimension of a much larger globalized food system that includes commodity markets, international trade, and complex distribution networks (Vermeulen, Campbell, and Ingram 2012). All told, lower agricultural productivity and yields may have enormous economic repercussions, including higher food prices, reduced agricultural employment, and out-migration of farm laborers to nearby cities, which in turn influences labor markets and wages in receiving areas. Changes in crop yields also have vital implications for food security as well as broader consequences for poverty rates and political stability. We return to these issues in chapter 8.

Climate change impact studies also examine the consequences of seasonal changes in climate. These may include shorter winters, an earlier arrival of spring, fewer winter nights with below freezing temperatures, or prolonged dry seasons. An example of changing seasonality can be observed in blossoming times for trees and flowers, which are becoming increasingly less predictable. As a result, pollinating insects are sometimes arriving in a region either too early or too late, creating mismatches between the emergence of flowers and the arrival of pollinators such as butterflies and bees (Hegland et al. 2009). While this might not seem like a big deal, close to 75% of all food crops rely, at least partly, on pollination by insects, birds, or mammals, and these seasonal mismatches can have substantial effects on crop productivity (Settele, Bishop, and Potts 2016). Changes in flowering dates of trees can also influence tourism and local economies. For example, in Washington, DC the annual Cherry Blossom Festival draws more than one million visitors from around the world, making significant contributions to the local economy. Changes in climate can mean that the festival (and the tourists) does not coincide with the actual blossoms. As one journalist wrote about the 2017 season in the *Washington Post*:

> We've had the most bizarre, semi-frozen and out-of-sync bloom at the Tidal Basin in recent memory, perhaps in the history of the cherry trees which were first planted in 1912. During the past four weeks, spring and winter have flip-flopped multiple times. Despite several freezes and a potent winter storm, a peak bloom did occur, about seven to 10 days later than first predicted. (Ambrose 2017)

Impacts studies sometimes investigate the effects of a combination of climatic conditions, such as increased temperatures *and* higher humidity, or increases in rainfall intensity *and* changes in wind patterns. Increases in the number of days with both very high heat and high humidity are a particular concern for human health, increasing the risk of heat exhaustion or heat stroke among sensitive populations such as the elderly, young children, and those who work outdoors (USGCRP 2016). The combination of heat and humidity also affects animal health, including that of livestock, which in turn affects production of meat, milk, and other food products (Rojas-Downing et al. 2017). Dairy cows, in particular, experience reduced milk production during very hot and humid weather (Wolfe et al. 2011).

Multiple stresses

While impact studies often focus on the direct and indirect effects of climate change, the integrative discourse reminds us that climate change is not occurring in isolation but is deeply intertwined with other large-scale processes of social and environmental change. As we have discussed in previous chapters, globalization, urbanization, deforestation, and land-use change are among many global-scale processes that are influencing the drivers of climate change. These global processes are also shaping the impacts of climate change, creating situations of **double exposure** in many sectors and regions (Leichenko and O'Brien 2008). For industries such as electronics and automotives, the globalization of supply chains means that the impacts of a climate shock on production facilities in one location are quickly transmitted to other locations (Liverman and Glasmeier 2014). For example, when Thailand experienced what was considered the worst flooding in fifty years in 2011, not only were production facilities damaged and more than 800 lives lost, but there were also cascading impacts in many other regions. Flooding of factories in Thailand that produced electronics, optics, automobile parts, and computer components meant that production facilities in Indonesia, Pakistan, Japan, the United States, and several other countries were shut down as key parts became unavailable (Perwaiz 2015). Double exposure also occurs in agriculture and other nature resource-based sectors where exposure to drought or other climate stresses may overlap with exposure to stresses associated with globalization such as market-based price fluctuations or changes in trade policies.

Teasing out the effects of double exposure to climate change and other large-scale processes of change is particularly complex when it comes to plants, animals, and ecosystems, many of which are already threatened by other human activities (Maxwell et al. 2016). For example, climate change is often discussed as a cause of future biodiversity and species extinction, yet biologists remind us that *baseline* extinction rates are already extremely high due to human activities. Current extinctions can be largely attributed to overexploitation, habitat losses associated with land-use changes (especially deforestation and the expansion of agriculture), invasive species, disease, and pollution (Maxwell et al. 2016). Exploitation of particular species for food, medicine, and other purposes has also had a significant impact. The demand for elephant ivory and rhino horns has, for example, threatened these populations in Africa. The introduction of new or invasive species (such as cats on islands, which devastate bird populations) has also played a role in extinctions. Finally, chemical pollution has threatened some species, especially as it accumulates in the food chain.

While a changing climate is affecting the abundance, distribution, and interactions among species, it is difficult to isolate climate change as the only cause of extinction because there are so many other threats. Nonetheless, extinction rates are projected to dramatically accelerate in the future as global temperatures rise. Under business-as-usual emissions scenarios, global extinction risks are projected to increase to 16% from current levels of 2.8%. In other words, one in six species is expected to go extinct with temperature increases of 4°C. This contrasts with a 5.2% global extinction rate associated with lowering emissions to a 2°C target (Urban 2015). Although the media often focuses on the effects of climate change on iconic species such as polar bears or on fascinating ecosystems such as coral reefs, climate change will affect the habitat and conditions for all types of living organisms.

Impacts of extreme climate events

For those of us who spend most of our time in climate-controlled homes, classrooms, and offices, or whose livelihoods are not directly dependent on the weather, awareness of the impacts of a changing climate often comes from television news or through social media coverage of extreme weather events, such as floods, heat waves, and wildfires. Yet extreme climate events are far more than trending topics on social media; they are also a major threat to human and non-human lives, livelihoods, property, ecosystem health, energy and transport infrastructure, and other assets. As we discussed in chapter 2, there is a broad scientific consensus that climate change loads the dice in favor of more extreme events (Hansen, Sato, and Ruedy 2012).

While not every extreme event can be attributed to climate change, evidence confirms that the frequency, intensity, and damage from extreme events are increasing worldwide (NAS 2016). In the case of the floods in northern England mentioned in the introduction, the rains were so exceptional that other common causes of flooding, such as land management, deforestation, or the draining of peat bogs were considered to have had little effect. The 2017 flooding in Houston from Hurricane Harvey was also exceptional. Although Houston's low-density development pattern contributed to the impacts, the sheer volume of rainfall in the region, estimated at more than 150 cm (60 inches) in some locations, was unprecedented (Blake and Zelinsky 2018). For both Manchester and Houston, climate change made these large precipitation events more likely, and it increased the amount of rainfall by a factor of 3.5 and possibly more (Risser and Wehner 2017; Otto et al. 2018).

In addition to extreme rainfall, many other recent weather events, including heat waves, have been shown to have a climate change fingerprint (Miao et al. 2016; Mitchell et al. 2016). For example, during the 2003 European heat wave, more than 22,000 heat-related human deaths occurred in northern Europe, including nearly 15,000 in France. Among those who died in France, the majority were elderly and female, and most lived in the Paris region (Pirard et al. 2005). Beyond loss of human life, the 2003 heat wave also contributed to the deaths of thousands of farm animals and a striking decline in agricultural production. Another example is the European and Asian heat wave that occurred in 2015. During July and August, even cities such as Wroclaw, Poland, where the maximum mid-summer temperatures are typically around 23°C (74°F), experienced an all-time high temperature of 38.9°C (102°F) (Erdman 2015). Heat waves often have severe consequences in regions that are not accustomed to sustained periods of very high temperatures. Extreme conditions also affected Europe, Southwest Asia and the Middle East in May and June 2017. The highest temperatures during this heat wave were experienced in Pakistan and Iran, where temperatures approached or exceeded 53°C, among the hottest surface temperatures ever recorded (Masters 2017). Yet even for regions accustomed to very high temperatures, climate change is expected to dramatically increase the frequency of extreme heat events. For example, extreme heat waves on the continent of Africa are projected to occur annually by 2045 under high-emissions scenarios (Russo et al. 2016).

While extreme heat is often associated with higher absolute temperatures in warm seasons, it may also entail unusually high temperatures during the winter. For example, in February 2018, recorded temperatures in the Arctic were 20°C (45°F) above normal (Samenow 2018). High winter temperatures in the Arctic reduce the amount of sea ice and snow cover, which has direct consequences for communities, and species that depend upon sea ice for hunting. Reduced Arctic sea ice and snow cover also has consequences for year-round weather patterns in other locations, including more frequent summer heat waves in mid-latitudes (Tang, Zhang, and Francis 2014). Another consequence of warmer winters is increased over-winter survival of many types of insects, including bark beetles, which pose risks to trees and forest ecosystems (Raffa et al. 2015). Warmer winter temperatures also affect agricultural production, particularly for fruit crops such as apples, which require cold winter temperatures for optimal spring bloom and fruit production (Wolfe et al. 2011).

With respect to extreme events, climate scientists have started to emphasize the importance of understanding **compound events** and the risks that they pose to society (IPCC 2012; Leonard et al. 2014). For example, the 2017 wildfires that burned across more than 400,000 hectares (1 million acres) in Northern and Southern California have been linked to a combination of strong winds, months of record-high temperatures, and long-term drought (Kahn and Mulkern 2017) (see Box 7.2). Other types of compound events are characterized by two or more extreme events occurring simultaneously or successively. The three tropical storms that hit the United States in September 2017 (Hurricanes Harvey, Irma, and Maria) could be considered one compound extreme event. In many cases, compound events are influenced by a combination of factors such as gradual sea-level rise, storm surges, and freshwater inflows from rainfall. In terms of impact assessment studies, the proliferation of compound events implies that a focus on a single indicator or event, such as extreme rainfall in one location, can underestimate the overall risks associated with climate change (Zscheischler and Seneviratne 2017).

Exposure to sea-level rise

When we think about the impacts of climate change, it is important to keep in mind that future conditions are not predetermined. There are many possible scenarios of future climates over the next century, which largely depend on processes and patterns of development, globalization, urbanization, and land and energy use. Given the difficulty of projecting specific impacts of climate change, there is increasing attention to assessing a range of possible impacts. Exposure analysis is widely used to identify which regions, sectors, or communities are most likely to experience rising sea levels and what rates of increase can be expected over short-, medium- and long-term horizons.

Concern about sea-level rise is not surprising. Coastal regions are already home to more than 40% of the world's human population. While all coastal zones will experience the effects of sea-level rise, areas that are particularly exposed include river delta regions, small island nations, and other low-lying settlements, where residences, businesses, and public infrastructure such as port facilities, roads, and bridges are located very close to current sea levels. These regions, sometimes described as **low elevation coastal zones** (LECZs),

Box 7.2 Climate change, wildfires, and forest ecosystems: Reverberating impacts

Climate change is increasing the risk of wildfires in many regions. Warmer temperatures and increased rainfall variability can extend the wildfire season in areas already prone to fire, such as Southern California in the United States, Spain and Portugal in Europe, and South Sumatra and Central Kalimantan in Indonesia. Climate change can also introduce new fire risks to areas that previously experienced none. Jolly et al. (2015) calculated that, as a result of climate change, the global mean fire-weather season increased in length by 18.7% from 1979 to 2013.

Warmer winters and drier conditions also contribute to fire risks by enabling insects such as the pine bark beetle to survive over winter, worsening tree infestations and increasing the severity of forest fires in weakened trees. Boreal forests in the northern hemisphere, found in northern Canada, Russia, Norway, and elsewhere, are particularly at risk because invasive insects are able to survive and thrive due to warmer winter temperatures (Raffa et al. 2015). The forest fires that nearly destroyed the city of Fort McMurray, Canada in 2016, for example, were caused by a combination of warmer and drier winter conditions and trees damaged by invasive insects (Gillis and Fountain 2016).

Forest fires are not only a threat to ecosystems and human settlements; they also have implications for human health as well as the global climate. Wildfire smoke is particularly dangerous because it contains particulate matter that gets into the lungs and blood, affecting humans as well as other species. Forest fires in Indonesia in 2015 are estimated to have contributed to 100,000 premature deaths from wildfire smoke (Koplitz et al. 2016). Because forests play a key role in the absorption of carbon dioxide globally, large-scale forest fires are also a concern with respect to the global climate system. In addition to the release of substantial amounts of CO_2 into the atmosphere, large fires create significant amounts of dark soot that is carried by winds and deposited in other places. When fires occur in northern boreal forests, some of this soot settles on ice sheets, thereby reducing their albedo and amplifying the warming effects, which contributes to increased rates of Arctic ice melt (Gillis and Fountain 2016). Along with a changing climate, expansion of suburban development into forested areas is also contributing to greater human exposure to fire risk in many regions.

Are forest or brush fires a risk where you live? Do you think that the risk of fire is something that you will take into account in making a decision about where to live in the future?

were home to more than 600 million people worldwide in 2000 (McGranahan, Balk, and Anderson 2007). By 2030, the size of the human population living in LECZs is expected to be close to a billion, with the largest exposed populations in China, India, Bangladesh, Indonesia, and Vietnam (Neumann et al. 2015). Many large and economically prominent cities are at risk from sea-level rise, including Guangzhou, New York, and Mumbai (Hallegatte et al. 2013).

Sea-level rise is ongoing, but it is occurring at different rates in different locations as the result of ocean circulation patterns, wind currents, groundwater management, and geologic conditions. Land areas in the US Pacific Northwest and Norway, for example, are still rising after the last glaciation, and this isostatic rebound effect partly offsets sea-level rise. By contrast, a number of large cities such as Bangkok, Manila, and San Francisco, as well as river delta zones such as the Mississippi Delta in the United States and the Ganges–Brahmaputra Delta in Bangladesh, are currently subsiding due to groundwater removal, diversion of rivers for irrigation and flood control, and other factors (Erkens and Sutanudjaja 2015). While estimates of future sea levels depend on projections of ocean temperatures and glacial melt, uncertainties about the rate of melting of the Greenland and Antarctic ice sheets complicate efforts to precisely quantify and project future exposure (Meyssignac et al. 2017). In order to convey the uncertainties associated with future sea-level rise, studies sometimes use a probabilistic approach and assign odds to different rates of sea-level rise (Kopp et al. 2017). The probability that sea levels will rise by 0.3 meters (1 ft) by 2100, for example, is much higher than the probability that they will rise 2.5 meters (8 ft) (see Table 7.1). Although there is a smaller probability of a 2.5-meter rise, this possibility cannot be dismissed, as outcomes are closely linked to the rate of melting of the Greenland and Antarctic ice sheets associated with different scenarios of temperature increase. Indeed, this is a very active area for scientific research and evidence indicates that rates of sea-level rise are accelerating (USGCRP 2017).

In considering the impacts of sea-level rise, it is important to recognize that even a small increase in sea levels means higher tides, greater storm surges, and more coastal erosion. In Arctic communities along the North Slope of Alaska, rising sea levels have contributed to substantial erosion in coastal areas, leading to damage to homes, roads, and infrastructure (Melvin et al. 2017). Record storm surges of more than 4 meters (13 ft) occurred in New York City and coastal New Jersey during Hurricane Sandy in 2012, as wind-driven water was pushed onto coastal lands. That surge flooded homes, businesses, and water supply and transport infrastructure, and also destroyed electric and gas supply lines, contributing to massive and long-lasting power outages (Leichenko et al. 2014). New York City's financial district was without power for more than a week, and some New York and New Jersey communities endured power outages for more than one month (Henry and

Table 7.1 Probability of exceeding global mean sea-level (GMSL) rise scenarios in 2100, based on IPCC Representative Concentration Pathways (RCP)

GMSL rise scenario	Low emissions RCP2.6	Medium emissions RCP4.5	High emissions RCP8.5
Low (0.3 m)	94%	98%	100%
Intermediate-Low (0.5 m)	49%	73%	96%
Intermediate (1.0 m)	2%	3%	17%
Intermediate-High (1.5 m)	0.4%	0.5%	1.3%
High (2.0 m)	0.1%	0.1%	0.3%
Extreme (2.5 m)	0.05%	0.05%	0.1%

Source: Sweet et al. 2017

Ramirez-Marquez 2016). In the aftermath of Hurricane Maria in 2017, 80% of Puerto Rico's electricity grid was damaged, destroyed, or compromised. The grid was particularly susceptible to damage from extreme weather events, due to decades of mismanagement, an economic recession, and failure to maintain and replace aging equipment (Gallucci 2018).

Information about climate impacts, extreme events, and sea-level exposure is vital for helping decision makers in the public and private sectors to understand how different regions may be affected by climate change and where policies and adaptive planning might be appropriate. In thinking about how this information is presented and conveyed, however, it is useful to recall our discussion in chapter 2 on the communication of scientific information. As we discussed, scientific tables, graphs, and reports are rarely enough to motivate action on climate change. One way to improve communication and reception of information about climate change impacts entails using personal narratives and stories to convey how climate change is affecting a local community (Jones and Peterson 2017). Accounts that people empathize with or connect to emotionally can make climate impacts more meaningful and relevant. For example, a coastal homeowner's description of how flooded roads from higher tides are making it difficult for him to visit his elderly mother who lives across town can be effective in conveying what the impacts of climate change mean for people's lives.

Another approach entails directly enrolling members of impacted communities in the development of climate assessments. This integrative method of research, sometimes referred to as **co-production** of scientific information, involves collaboration between community members and scientists to identify assets or activities that they feel are valuable and might be at risk (Kruk et al. 2017; Bremer and Meisch 2017). Ensuring that such engagement is inclusive of a broad range of community members, including representatives of socially marginalized groups, is a key element of the co-production process. As we show in later chapters, inclusive approaches to climate change responses are also vital for promoting successful adaptation and equitable transformations to sustainability.

Cultural impacts of climate change

The impacts of climate change are often portrayed as a threat to entire cultures and ways of life, particularly for communities living in small island states in the Pacific or Caribbean, Arctic regions, and the Bangladesh Delta. Yet the idea that cultures can simply be wiped out by climate change has been criticized and contested on various grounds. Strauss (2012) emphasizes that cultures are not fixed but are instead learned, shared, and always changing, with cultural change being a key driver of adaptation. Coastal communities confronting sea-level rise on the Pacific island of Tuvalu, for example, consider themselves to be strong and self-reliant rather than vulnerable victims; population mobility is an integral part of their past and present (Farbotko and Lazrus 2012).

While the relationship between climate change and culture is indeed complex, there is no doubt that there are and will be cultural losses – both material and non-material – from climate change (Adger et al. 2012). Some of the most important cultural impacts of climate change are related to experiences that are valued by communities and groups. Whereas climate impact assessments may project how specific variables will change, they tell us very little about how cultures will experience these changes. For example, projecting how much snow will fall in a particular region in the future under various greenhouse gas emissions scenarios can provide critical information about local hydrology or revenue projections for a ski resort, but it says little about how cultural values will be affected (i.e., how communities are likely to experience barren mountain slopes, missed opportunities for traditional winter sports, or unprecedented winter flooding). Similarly, while models of sea-level rise may produce maps of changing coastlines, they provide little insight into how particular communities will experience damage to or loss of fisheries, beaches, sacred sites, monuments, or species that are located in exposed coastal areas (Brady 2018).

As seasonal rhythms shift and species change their distributions and behaviors, traditional environmental signals that are critical to human activities also become less consistent and reliable. Although traditional ecological knowledge is continually updated, the variability and uncertainty of climate change make it increasingly difficult to read, interpret, and project conditions that are considered important to livelihoods and safety. Loss of humanity's biocultural heritage means loss of different types of knowledge about the natural world and its relationship to humans (Gavin et al. 2015). Such knowledge is considered vital to a community's cultural values and heritage, and its loss has both material and spiritual consequences (Adger et al. 2014). The loss of glaciers in the high mountain regions of Peru, Nepal, and China, for example, is influencing the way that local people make meaning of their environment and their sense of identity (Allison 2015). Interviewing residents confronting sea-level rise along the Mississippi Delta, Sack found that **place attachment** was critical for their sense of community and connection:

> [M]any cannot imagine living anywhere else. They complain that the bloodless cost–benefit formulas determining which communities get protection give little weight to the qualities that make Lafitte desirable . . . The policymakers "don't place value on anything but the money, not the longevity of these communities, not the culture," said Tracy Kuhns, 64, a longtime resident of the Barataria community across the bayou from Jean Lafitte. "One of the problems in this country is that people don't have any connection to where they live. People really want that. Why would you take it away from people who already have it?" (Sack 2018)

Many places that are valued as part of humanity's cultural heritage are at risk from rising sea levels and other changes. This includes iconic cultural landscapes, structures, parks, and gardens (Hall et al. 2016). It also includes places that hold sacred, symbolic, and spiritual meanings for particular cultures, such as temples, burial sites, or locations that are significant for rituals (Kim 2011; Allison 2015). Studies of the impacts of climate change on world heritage sites and cultural icons have identified threats to structures, monuments, and landforms from changes in moisture, humidity, aridity, wind, and weathering processes. In a study of the potential consequences of sea-level rise for the 720 World Heritage Sites identified by UNESCO, Marzeion and Levermann (2014) found that under different scenarios of climate change, anywhere between 6.5% and 20.7% of all sites would be affected. In addition to direct impacts on cultural heritage sites, climate change will also influence heritage tourism and the livelihoods associated with it (Hall et al. 2016). As such, climate change is amplifying existing economic, institutional, and development challenges associated with managing and maintaining these sites (Markham et al. 2016).

There are many ways that climate change will harm culturally valued assets, activities, and ways of life, but the linkages between climate change and culture are also multidirectional (Hulme 2015). Cultural change will likely be an essential part of climate change responses and transformations towards an equitable and sustainable world. Discussing the impacts of climate change opens reflection and dialogue about what is valued and why, what type of future is considered desirable, and, not least, who or what determines present and future outcomes. It also points to the types of knowledge that are needed to preserve both cultural and biological diversity. Diverse sets of knowledge, together with new knowledge about the implications of climate change for social and natural systems, may prove to be powerful catalysts for both preserving and transforming cultures.

Summary

Climate change is transforming human and natural systems worldwide. This chapter has examined the tangible and intangible impacts of climate change for sectors, ecosystems, regions, communities, and cultures. Taken as a whole, this detailed picture of the consequences of a changing climate makes a powerful case for action to reduce these risks. In the next chapter, we consider social, economic, and political factors shaping these outcomes as we explore what climate change means for broader questions of vulnerability and human security.

Reflection questions

1 One big concern about climate change is related to the irreversibility of impacts. What types of impacts do you think of as irreversible, and why?
2 We live in a society where people deal with risks every day (for example, when we travel or drive). Yet many people find that climate change risks are more difficult to take seriously than other types of risk. What explains this?
3 Are there any aspects of your culture threatened by climate change? Which cultural values do you consider most important to maintain?

Further reading

Adger, W. N., Barnett, J., Brown, K., Marshall, N., and O'Brien, K., 2012. Cultural dimensions of climate change impacts and adaptation. *Nature Climate Change* 3(2): 112–17.

Hallegatte, S., Green, C., Nicholls, R. J., and Corfee-Morlot, J., 2013. Future flood losses in major coastal cities. *Nature Climate Change* 3(9): 802–6.

National Academies of Sciences, Engineering, and Medicine (NAS), 2016. *Attribution of Extreme Weather Events in the Context of Climate Change.* Washington, DC: The National Academies.

Strauss, S., 2012. Are cultures endangered by climate change? Yes, but . . . *Wiley Interdisciplinary Reviews: Climate Change* 3(4): 371–7.

Vermeulen, S. J., Campbell, B. M., and Ingram, J. S. I., 2012. Climate change and food systems. *Annual Review of Environment and Resources* 37(1): 195–222.

8 VULNERABILITY AND HUMAN SECURITY

Climate change as a threat

Have you ever moved to a new neighborhood, to a new town or city, or even to a new country? What was it like to make this move? Have you been able to return to visit your old home? Usually a move elicits mixed feelings. You may look forward to living in a new place but also feel sad to leave behind friends and family. Yet *what if you had to move* because the land you lived on was slowly disappearing under rising seas or because your community was running out of fresh water? While these might sound like remote threats or even plotlines from a dystopian novel, they are in fact real dangers that many individuals and communities are currently facing. Sea-level rise is threatening homes and communities in low-lying island countries such as Micronesia, the Maldives, and Tuvalu, and along river deltas and coastlines in many nations. Water shortages are undermining livelihoods for Turkana communities in northwestern Kenya and disrupting ways of life in cities such as São Paulo, Brazil and Cape Town, South Africa (Maxmen 2018; Sengupta 2018). In many other locations, the impacts of climate change are making it difficult for individuals, households, and communities to meet their basic needs for food, water, and shelter, and to safeguard health and well-being.

In this chapter, we examine the implications of climate change for vulnerability and human security. We begin with the concept of vulnerability, showing how susceptibility of individuals and communities to harm from climate change is shaped by social, economic, political, and institutional contexts. We then consider the implications of climate change for human security. We explain how climate change can contribute to food, water, and health insecurities, particularly for vulnerable populations that are burdened by poverty or face other social, economic, political, or environmental constraints. We conclude by drawing connections between climate change and broader security concerns, exploring whether and how climate change may act as a **threat multiplier** that contributes to conflict and migration.

Vulnerability

The concept of vulnerability draws attention to social, economic, and political factors that make particular nations, communities, individuals, and groups more sensitive to harm from climate change. At the national level, vulnerability to climate change is often said to mirror, in reverse, relative contributions to greenhouse emissions. Those countries that are contributing least to emissions are highly vulnerable to harm from its impacts, while those contributing most tend to have a greater capacity to adapt (see Figure 8.1). Although a focus on national-level vulnerability brings out the international equity issues associated with climate change impacts, it can also hide differences in vulnerability to climate change within countries and regions, and even among cities, communities, groups, or households (O'Brien, Sygna, and Haugen 2004).

Defining and assessing vulnerability

Vulnerability to climate change can be defined as the predisposition or likelihood of being adversely affected by a climatic event or circumstance (Adger 2006; Eakin and Luers 2006). Vulnerability can be interpreted in different ways, depending on the discourse and context (O'Brien et al. 2007). Within the biophysical discourse, vulnerability is closely associated with physical exposure, sensitivity, and impacts of climate shocks and stresses such as heat waves, droughts, and sea-level rise, as discussed in the last chapter.

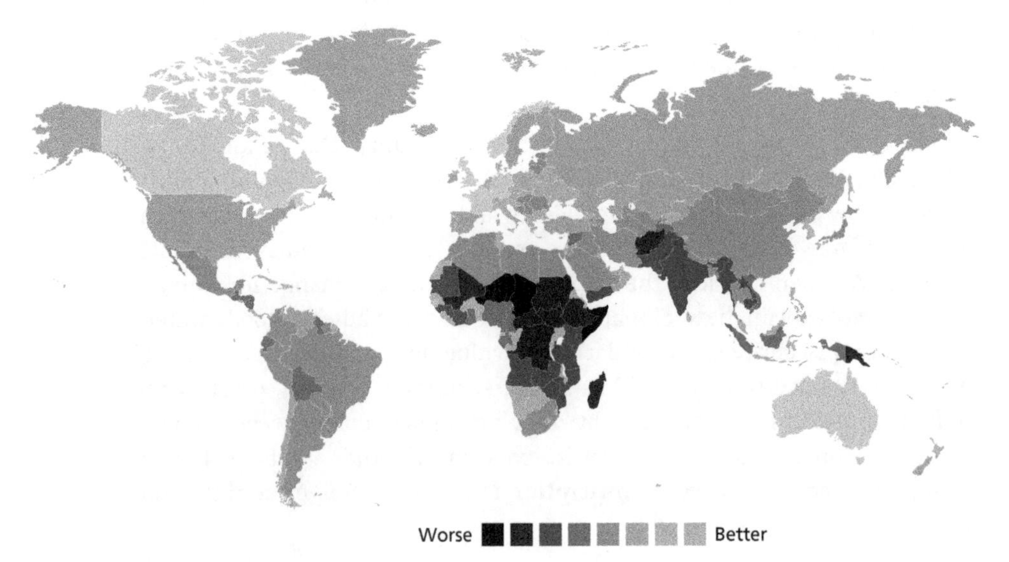

Figure 8.1 National vulnerability map.
Source: Based on the ND-GAIN Country Index 2016. Available at https://gain.nd.edu/our-work/country-index/

Critical discourses on vulnerability emphasize the economic, political, and institutional circumstances and conditions that make some individuals, households, and groups more sensitive to harm from climate events or less able to respond and cope with changes when they occur.

As illustrated in Figure 8.2, vulnerability results from the interaction between exposure to climate change and a variety of factors that influence both sensitivity to harm and capacity to adapt. Within a flood-prone region, for example, all households may be exposed to a flood event but sensitivity and **adaptive capacity** depend on many other factors. Some households may be relatively well prepared for flood events. Their houses may have been elevated to sit well above common flood heights, thus reducing their exposure. They may have also purchased flood insurance to cover the risk of damage. Other households in the same community or region may be far less prepared. They may not be able to afford to elevate or otherwise protect their homes from flood exposure, may not have access to affordable flood insurance, and may not be mobile if evacuation is necessary. The latter group is considered to be more vulnerable to floods, but it is not the storm event alone that created these conditions.

Vulnerability assessments can help to clarify the complex mix of factors that explain why particular households, groups, and regions are more prone to harm or less able to respond to climate stresses (see Box 8.1). Social factors such as income and wealth, education, age, health status, and physical disability may influence sensitivity to climate stresses and ability to respond or plan for climate extremes. In the case of heat waves, for example, the elderly and very young, and those with existing health conditions such as asthma, are typically more sensitive to the effects of extreme heat. This vulnerability is compounded by social factors such as living alone, lacking an effective social network, or an absence of early warning systems. There is also a psychological dimension of vulnerability that recognizes that individuals will react and respond differently to climate shocks and stresses (Grothmann and Patt 2005). Psychological factors, earlier life experiences, and prior traumas that may be triggered or reinforced by extreme events can make

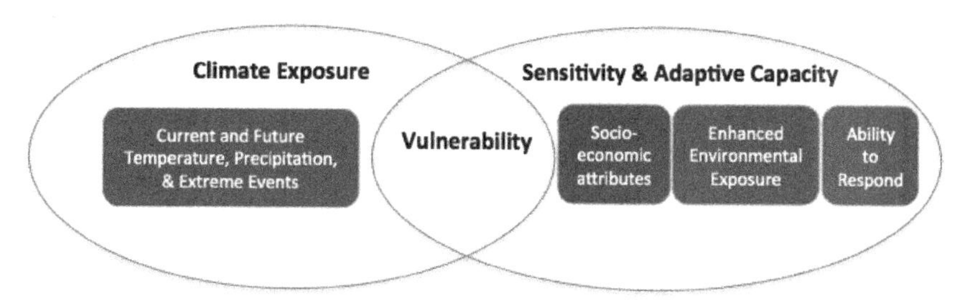

Figure 8.2 Vulnerability diagram.
Source: Petersen et al. 2014. Adaptation International (www.adaptationinternational.com).

Box 8.1 Mapping spatial vulnerabilities

Spatial vulnerability assessments entail mapping of vulnerability patterns within a city, nation, or other geographic region. Drawing from hazards research (Cutter, Boruff, and Shirley 2003), vulnerability mapping typically involves identification of factors that influence exposure to climate change shocks and stresses, susceptibility to harm, and capacity to adapt and recover. These assessments are often used to identify **vulnerability hot spots** where resources and policies should be targeted to reduce risks.

Within a spatial vulnerability map, social factors such as poverty levels, unemployment rates, levels of educational attainment, gender equality, and other factors are often used to pinpoint differences in vulnerability among communities and regions. Figure 8.3 depicts differences in social capacity to respond to climate change across districts in

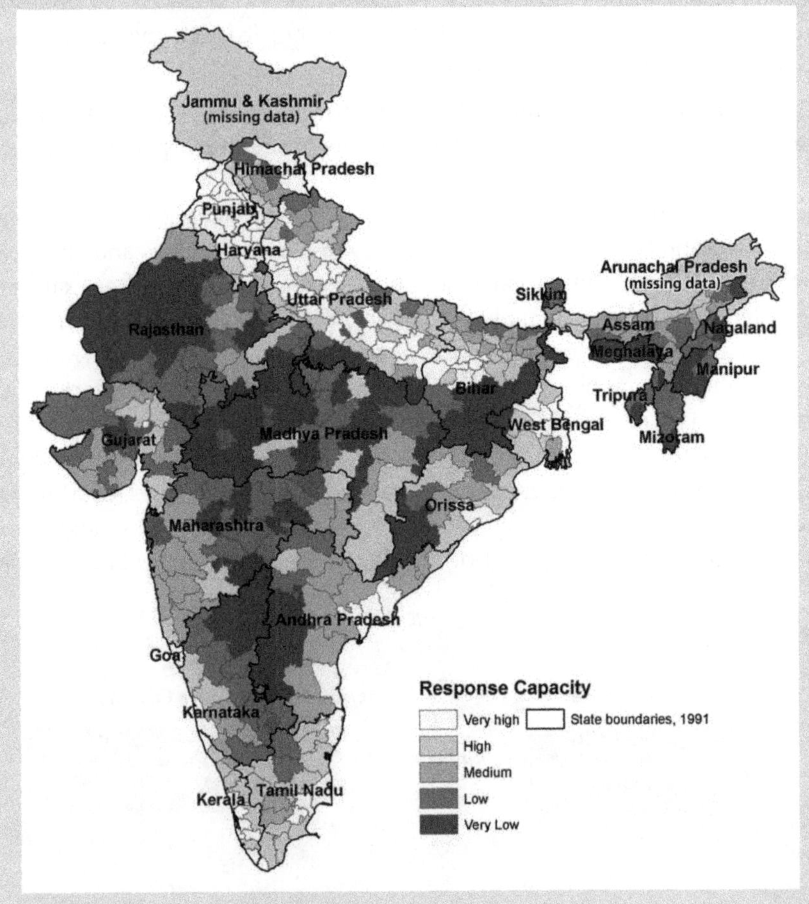

Figure 8.3 Map of response capacity in districts in India.
Source: Leichenko and O'Brien 2008

India based on a composite index of social factors. The figure depicts higher degrees of response capacity in most districts located along the Indo-Gangetic Plains in northeastern India and lower response capacity in the central and western interior portions of the country (Leichenko and O'Brien 2008).

If you were to map social vulnerability in your city or region, what kinds of indicators would you use? Which climate stresses are most important where you live? What factors influence the social vulnerability of individuals and groups in your region?

certain individuals or groups more susceptible to harm from climate shocks or less able to respond.

In addition to attributes that vary among individuals and households, vulnerabilities can also be traced to weaknesses or failures in social networks, institutions, and governance, including the presence (or absence) of early warning systems, health care systems, or insurance. Weak governing institutions may mean that there are few resources and little capacity to invest in early warning systems or programs to aid in post-event recovery. In studying vulnerability to heat waves in communities in Sydney, Australia, for example, Zografos, Anguelovski, and Grigorova (2016) found that denial of climate change and absence of government support for heat-wave warning campaigns contributed to a lack of community adaptive capacity. Limited public awareness of heat-related dangers, including low perception of risk and little prior experience with heat waves, were also found to contribute to the community's vulnerability.

Vulnerability can also stem from how people make a living. Those who work outdoors or whose livelihoods and jobs depend on climate-sensitive industries such as agriculture, lumber and forestry products, fisheries, tourism, and outdoor recreation are more likely to be directly exposed to climate shocks and stresses. While changes in temperature and precipitation regimes can have direct effects on production, output, or harvest, vulnerability also depends on the characteristics of the affected job or livelihoods. Subsistence farmers, hourly workers, workers in the informal economy, and small purveyors of all types are typically often more vulnerable because of a limited ability to protect themselves and fewer assets to use for recovery from climate shocks (Leichenko et al. 2014).

Vulnerability assessments also recognize the influence of double exposure to other large-scale, global processes such as globalization, urbanization, and land-use changes, which can affect both susceptibility to harm from climate shocks and stresses and capacity to respond (Leichenko and O'Brien 2008). For farmers, fishermen, loggers, and tourism operators, climate-related disruptions to their livelihoods often coincide with changes in global market conditions as a result of both rising competition from new producers and production sites or changing consumer preferences. In less industrialized cities, rapid population growth and consequent expansion of informal housing settlements is enhancing the exposure of many residents to flooding, mudslides, and other climate stresses. However, the vulnerability of exposed groups is also shaped by other contextual factors. Housing market dynamics often mean that informal or slum housing is the only type of shelter that poorer residents and rural in-migrants are able to afford. In the case of Nairobi, for example, about 60% of the population lives in informal settlements that occupy only 6% of the land in the city (Davis 2006). Additional factors such as political patronage or corruption can sometimes mean that particular groups or areas are prioritized over others following an extreme event, influencing the speed and extent of recovery efforts.

Vulnerability, inequality, and intersectionality

Vulnerability to climate change may also be shaped by inequalities associated with gender, income, caste, race, or ethnicity. For example, gender-related vulnerabilities within particular societies are influenced by uneven power relations, different rights and responsibilities, and unequal access to resources and land (Nightingale 2011). In identifying vulnerable groups, it is important to recognize the **intersectionality** between different forms of discrimination and economic disadvantage. In the city of Lagos, Nigeria, low-income women were found to experience more severe impacts during a 2011 flood event than middle- or high-income women. Greater vulnerability of low-income women in this case stemmed from the intersection of poverty, unsafe housing, and loss of livelihoods with gender-role expectations around childcare and household duties, and lack of access to health care (Ajibade, McBean, and Bezner-Kerr 2013).

Although awareness of the connections between vulnerability and other forms of discrimination often center on less-industrialized countries, poorer and racially or ethnically marginalized populations in industrialized countries may also be disproportionately vulnerable. In these contexts, vulnerability of poorer populations is sometimes tied to infrastructure and the provision of public services. For example, climate-related disruptions of urban public transport in cities such as New York and Philadelphia can have disproportionate effects on poorer, minority populations, who are less likely to have alternative transport options during weather-related shutdowns (Barnes 2015). Lack of access to support services and social networks can also contribute to the vulnerability of poor individuals, particularly those who are elderly and isolated (Klinenberg 2015).

While inequality and other social conditions are sometimes seen as "causes" of vulnerability, the critical discourse draws attention to root causes, including historical contexts and the power of particular groups with vested interests in maintaining inequalities (Ribot 2014). Emilie Cameron (2012) points to the role of colonial, postcolonial, and decolonizing histories, practices, and ideas in creating and perpetuating vulnerability. Her research on indigenous communities in the Canadian Arctic highlights the ways that vulnerability assessments themselves can facilitate external control and resource exploitation, perpetuating paternalistic and exploitative approaches towards these groups. Labeling a region or group as "vulnerable" can also mask existing capacities and potentials and direct attention away from the social, economic, and political factors that contribute to vulnerability (Barnett, Lambert, and Fry 2008). Attention to biophysical characteristics, such as semi-arid conditions or low-lying terrains, may also obscure the social dimensions of risk and vulnerability (Barnett and Campbell 2010). Labeling some locations and populations as vulnerable can also lead to a sense of complacency among those who are not categorized as such. This is particularly the case when people assume that they will not be affected and hence do not take actions to reduce risk and vulnerability (O'Brien et al. 2006).

Human security

Like vulnerability, security is a concept with many meanings and interpretations. While national and international security discussions tend to focus on the geopolitical interests and concerns of nation-states, the notion of human security emphasizes the well-being and rights of individuals and communities. Human security includes having the resources, capacity, and voice to influence conditions that affect the well-being of one's self, family, or community (O'Brien and Barnett 2013) (see Box 8.2). It is a normative concept with aspirational dimensions; it acknowledges that many people currently live under conditions of *insecurity* and recognizes that these conditions are not natural or inevitable, but instead they are closely related to social, political, and institutional contexts (O'Brien and Barnett 2013).

A focus on human security builds upon our discussion of vulnerability by drawing explicit attention to the avenues through which climate change directly or indirectly harms vulnerable individuals and communities and diminishes their capacity to respond to all types of threats and stresses. Threats to human security are often linked to climate-related impacts that directly influence lives and livelihoods. However, human security also recognizes the role of political, social, and economic processes that shape individual and collective capacities to cope, adapt, and respond. Living in the presence of political instability, failed governments, and violent conflict often makes it challenging to meet human needs for food, water, shelter, energy,

Box 8.2 Human security and human rights

Human security can be defined as "a condition in which people and communities have the capacity to respond to threats to their basic needs and rights, so that they can live with dignity" (O'Brien and Barnett 2013: 375). This definition recognizes the universal rights of individuals and communities. These are rights that have historically been denied to many and are now challenged by new threats and risks, including climate change. Appearing in the aftermath of World War II, the concept initially emphasized safety from chronic threats and disruptions, including war and hunger, or "freedom from fear and freedom from want" (Liotta and Owen 2006).

Human security is a progressive idea that developed alongside conceptions of human rights, including the 1948 Universal Declaration of Human Rights, which recognizes that "All human beings are born free and equal in dignity and rights" and that "Everyone has the right to life, liberty and security of person." In 1994, the United Nations Development Programme (1994) identified seven dimensions of human security: personal, environmental, economic, political, community, health, and food security. The Commission on Human Security (2003), led by Nobel Prizewinning economist Amartya Sen, later refined the definition to include core values: "[t]he objective of human security is to safeguard the vital core of all human lives from critical pervasive threats, in a way that is consistent with long-term human fulfillment" (Alkire 2003: 2).

Do you think these definitions of human security are still relevant today? Are there any other dimensions of human security that you think should be included in the definition?

and health. Such conditions can also contribute to emotional and psychological trauma and the breakdown of social networks. As an integrative concept, human security recognizes both external risks and the intangible and subjective perceptions of threats that can undermine human well-being.

Food insecurity

Food security is a critical element of human security. According to the Food and Agriculture Organization of the United Nations (FAO), world hunger has been increasing since 2014, with the number of undernourished people reaching 815 million in 2016 (FAO 2018). Globally, the vast majority of food-insecure individuals live in sub-Saharan Africa and South Asia. At the same time, more than 600 million adults worldwide are considered to be obese, and the share of the world's adult population that is obese is growing. In high-income and upper-middle-income countries, obesity and food insecurity often coexist in poor communities and households, where individuals are consuming excess calories but their food lacks adequate nutrients. The twin crises of undernourishment and obesity are stark reminders that food security is not simply about consuming enough calories but also about having access to safe, nutritious, and affordable foods (see Figure 8.4).

Climate change may exacerbate food insecurities through a number of pathways linked to impacts on agriculture, fisheries, and livestock production (Myers et al. 2017). Food insecurities are already a chronic concern for small-scale producers in regions that are drought prone and dependent on rain-fed agriculture, and where structural factors limit food availability, access, and utilization. Climate change directly threatens the food security of these farmers, whose main source of nutrition comes from food they produce for themselves. In times of shortage, subsistence producers must obtain more of their food from markets, which requires that they have sufficient money or can use other means of exchange, such as trade, barter, vouchers, or gifts.

In addition to the direct threats of climate change to food production, food insecurity can also be triggered by climate-related price shocks. Higher food prices as a result of drought or other climate stresses can create food insecurity for people who are already living in precarious economic situations and who spend a significant share of their income on food. In both rural and urban

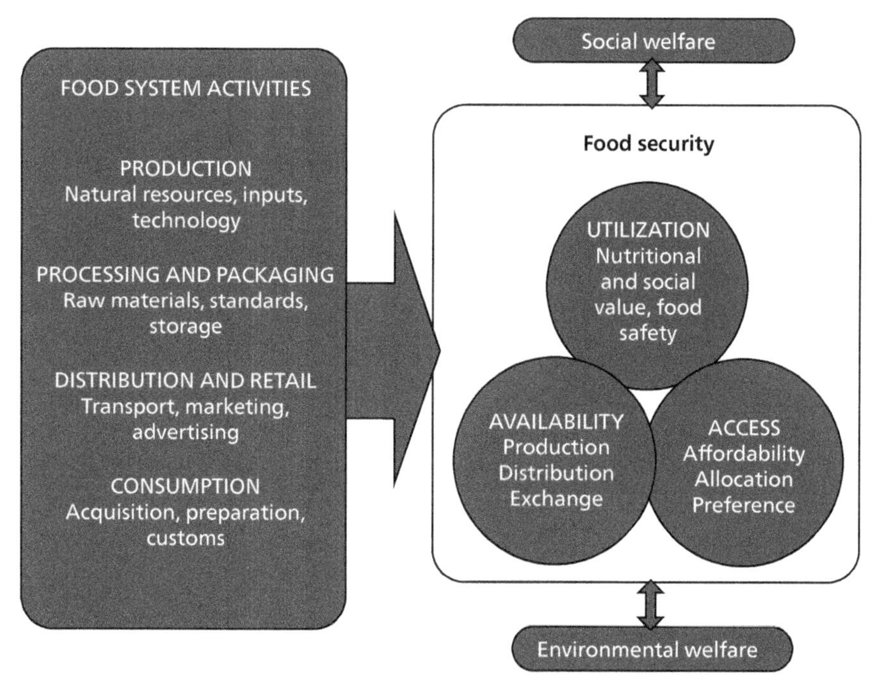

Figure 8.4 Food security.
Source: Sandstrom and Juhola 2017

areas, price shocks may mean that some members of a household either go without food or reduce the number of meals consumed. Rising food prices have the most acute effects on food-insecure populations in lower-income countries, but they also contribute to food insecurity within middle- and high-income countries. Although social safety nets in wealthier countries mean that fewer people will go hungry, affected individuals may nonetheless shift consumption to cheaper but lower-quality foods (Edwards et al. 2011).

The implications of climate change for food security are also shaped by the politics and economics of food access and consumption in different communities and regions (Wheeler and von Braun 2013). This includes uneven power relations which influence control over the land, water, seeds, and biological diversity necessary to produce food (Jarosz 2014). Of particular concern in some regions of Africa, Latin America, and Southeast Asia is the phenomenon of **land grabbing**, which refers to increased external control of land ownership and land use (Daniel 2011; Rulli, Saviori, and D'Odorico 2013). This is not a new phenomenon, but in recent years there has been an expansion in the scale and extent of land ownership by foreigners (Zoomers 2010). Land grabbing can contribute to vulnerability and food insecurity by promoting activities such as production of biofuels, expansion of irrigated crop production for export, and construction of large-scale tourist complexes, many of which place demands on water supplies and other resources needed for the production of local food crops.

In the future, climate change is likely to influence food production at local, national, and global scales. As discussed in earlier chapters, climate-related shifts in production zones, changes in pollination biology, increases in pests and disease, and changes to marine ecosystems together will affect global food supplies. In terms of food security, the effects of these combined pressures will be highly uneven among different population groups, with the poorest bearing the greatest burden (IPCC 2014b). At the same time, trends in both population growth and consumption suggest that global demand for food, including grains, meat, fish, and other proteins, will increase until the middle of the twenty-first century (Godfray et al. 2010). Feeding an expected population of nine billion people will place tremendous pressures on land, water, and energy, and these pressures will be exacerbated by climate change. Food security in the future will be influenced by decisions taken today that reduce future risks and vulnerabilities, as well as by economic, political, institutional, and cultural factors that shape agricultural practices and dietary choices.

Water insecurity

Availability of fresh water is another facet of human security that will be influenced by climate change. Globally, more than two billion people currently lack access to safe water supplies. While water insecurity is present in nearly every region of the world, it is especially acute in areas of sub-Saharan Africa, South Asia, and the Middle East (Gain, Giupponi, and Wada 2016). Climate change adds to pressures on water availability and quality by changing the timing and predictability of surface water supplies and increasing the frequency and magnitude of both droughts and extreme rainfall events. However, as is the case with food security, water security is not just about the overall supply of water – it is also about social factors that influence whether and which individuals and households have access to sufficient quantities of clean water (Gain, Giupponi, and Wada 2016). Governance of water supplies, rules and norms about water access, control and management of water infrastructure, and implementation of market-based pricing policies all have consequences for water security (Leichenko and O'Brien 2008).

While climate change can affect availability of water for entire regions, cities, and communities, the consequences of these changes often mirror existing and historical patterns of inequality and vulnerability that are linked to the distribution of political and economic power. In the case of Cape Town, for example, water shortages in 2018 were influenced by climatic conditions, but they were also a reflection of longer-term water management challenges for the municipality, including adapting to growing demands and competition with agricultural uses (Ziervogel, Shale, and Du 2010). In terms of impacts, the shortages posed a disproportionate burden on poor residents living in informal settlements. The majority of these settlements are located in a low-lying, flood-prone area of the city known as Cape Flats (Ziervogel, Waddell, et al. 2016). In response to the shortages, many residents of informal settlements had to both ration limited water supplies and travel significant distances to collect water from designated municipal collection points. By contrast, many of the city's wealthier residents were able to respond through actions such as purchasing water from private sources, hiring contractors to dig private wells, or building water tanks (Sieff 2018).

Water insecurities also have a strong gender dimension. In many cases, women's experience of water insecurity can be explained by cultural norms and expectations about water use, adequacy of supply, and constraints on physical access (Stevenson et al. 2012: 393). For example, in rural areas of South Asia and sub-Saharan Africa, women in poor households are typically responsible for collecting water from outdoor boreholes or traditional wells to meet their family's needs. Droughts can make this task more difficult and time-consuming by drying up nearby sources. These conditions can also place women and girls at increased risk of sexual violence because they are traveling farther from home to collect water and spending more time in unfamiliar surroundings far from their own communities (CARE 2016). Commodification of water within a market system may also contribute to water insecurity for women. Discussing the gendered aspects of water security among farmers in rural Australia, Alston (2011: 61) found that "men are far more likely to have access to economic resources than women, and hence to control the way water is used and distributed." Ensuring water security for vulnerable populations in the face of climate change thus requires attention to both physical factors that influence the provision of water supplies and to other political, economic, social, and cultural processes that shape water access and consumption.

Health and well-being

Another human security concern associated with climate change relates to health and well-being. Health is not merely the absence of disease. It has been defined more broadly by the World Health Organization as "a state of complete physical, mental and social well-being" (WHO 2018). Climate change will influence many aspects of human health (see Figure 8.5). Beyond immediate dangers of weather-related illness and fatalities, such as heat stroke or heat exhaustion, many additional facets of human health are likely to be influenced by climate change. These include both direct physical threats, such as vector- and water-borne disease, as well as threats to mental well-being.

Disease threats

Susceptibility to vector-borne diseases, such as malaria, dengue, and yellow fever, is often highly intertwined with other threats to human security (Wu et al. 2016). These diseases, which are transmitted by insects such as mosquitos and ticks, disproportionately affect populations living in poorer countries in Africa, South and Southeast Asia, and South America (see Figure 8.6). In regions where these diseases are endemic, populations that are poor, malnourished, and lack access to sufficient and safe water are most susceptible to illness, with pregnant women, infants, and children typically facing the highest risks (Huynen, Martens, and Akin 2013).

Warmer temperatures and more variable rainfall as a result of climate change are expected to increase the risk of exposure to vector-borne diseases by expanding suitable habitat for ticks and mosquitos that carry these diseases. However, the transmission and spread of these diseases is a complex phenomenon that depends on interactions between ecological, social-economic, and institutional factors, among others (Huynen, Martens, and Akin 2013; Parham et al. 2015). Increasing exposure to dengue and

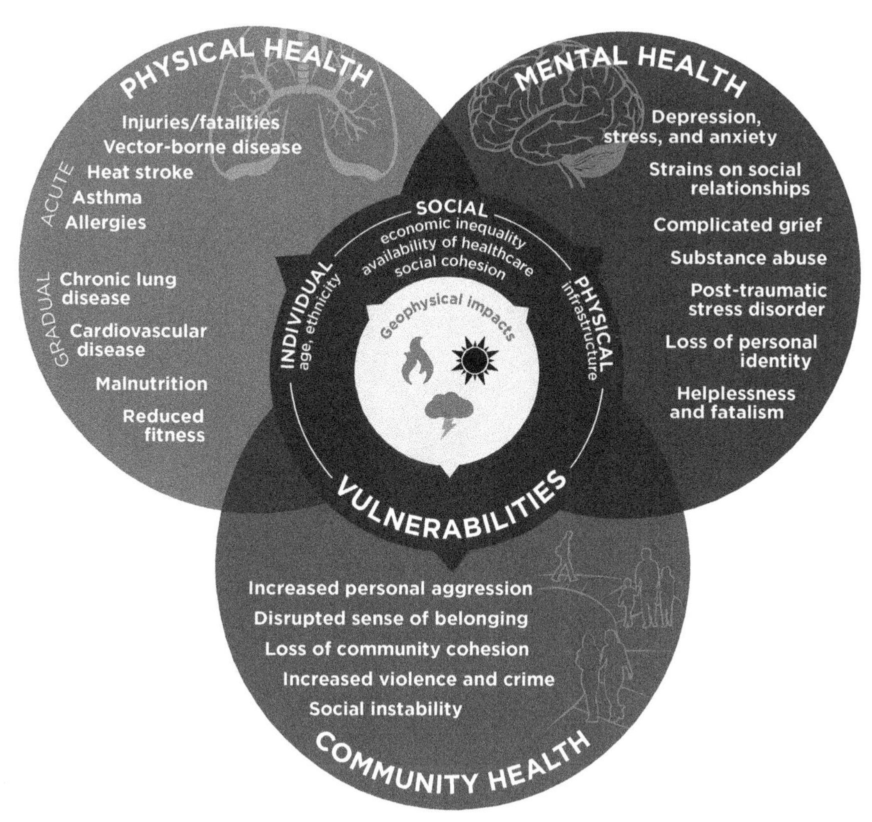

Figure 8.5 Climate change and health.
Source: Clayton et al. 2017. © American Psychological Association, and eco America.

malaria, for example, is a particular concern for populations living in environments where breeding grounds for mosquitos are plentiful, and where there are limited resources or capacity for mosquito-control efforts. This includes informal urban settlements, areas disrupted by floods or other hazards such as earthquakes, and zones of conflict or political instability. In addition to vector-borne diseases, climate change also increases risks of flood-related contamination of drinking water supplies, leading to greater exposure of vulnerable populations to waterborne diseases like cholera, as well as many other viral, bacterial, and parasitic illnesses (Hunter 2003).

Greater disease exposure as a result of climate change is also becoming a concern in temperate and colder regions. Within North America and northern Europe, the distribution of disease-carrying ticks is shifting north and to higher elevations (Mysterud et al. 2016). Together with changes in land use, deer populations, and other factors, these shifts have led to an increase in human contact with infected ticks, increasing rates of Lyme disease and tick-borne encephalitis in many locations. Within the United States, reported

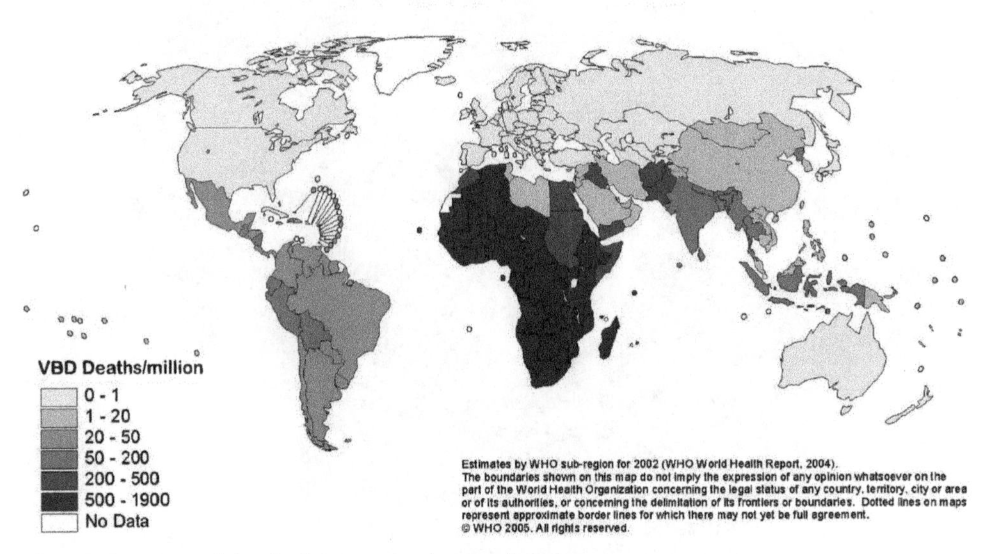

Figure 8.6 Map of deaths from vector-borne disease.
Source: WHO 2005

cases of vector-borne diseases from ticks and mosquitos more than doubled between 2004 and 2016 (Rosenberg 2018).

Beyond the direct threats of contracting disease, climate change may also undermine the ability to treat many types of diseases. Both traditional and conventional medicine are derived from many types of plants. Indeed, it is estimated that over half of the world's population relies on medicinal plants (Wong 2001). The pharmaceutical industry also derives many widely used medications from compounds originating in plants. Along with other threats to biodiversity described in chapter 7, climate change will influence the habitats and microclimates of many plants, which can affect their abundance, distribution, and, ultimately, their continued existence.

Psychological, emotional, and mental health consequences

While the measurable impacts of climate change on food, water, livelihoods, and health are well documented, the psychological and emotional consequences of climate change are also increasingly recognized. The impacts of climate change on mental health typically depend on both exposure to a climate shock or stress and the social, cultural, and historical contexts in which climate-related disruptions are experienced (Fritze et al. 2008; Swim et al. 2011). Extreme climate events in particular can take a significant toll on mental health, whether through the experience of losing one's home or possessions or being in physical danger due to a storm, flood, or wildfire. Mental health impacts of climate change may also stem from sadness, grief, or solastalgia associated with an ongoing or impending loss of loved species, locations, or experiences (Cunsolo and Ellis 2018). Such disruptions may be experienced through emotions such as anxiety, stress, trauma, social isolation, and depression. These and other psychological aspects of climate change can have tangible socioeconomic and physical consequences, whether through missed work or school, a reduced capacity to concentrate or participate in everyday activities, or an increased consumption of medicines or other healing or pain-reducing modalities. In some cases, climate change may contribute to increased rates of domestic violence or attempted suicide (Fritze et al. 2008).

Acute stress typically occurs for all population groups during catastrophic events such as hurricanes, floods, or wildfires and diminishes after safety and security have been reestablished (Fritze et al. 2008). Some of the factors that make people more susceptible to longer-lasting mental health impacts include age, preexisting disabilities, chronic illnesses, socioeconomic and demographic inequalities, income, and education levels (Clayton et al. 2017). Worldviews have also been shown to play an important role in emotional and psychological responses to shocks and stresses (Edmondson et al. 2011). Children are considered particularly vulnerable to experiencing psychological harm, as climate change-related traumas can impede healthy physical, social, mental, and emotional development. In the aftermath of disasters, children often require different types of support than adults (Peek 2008). For some individuals, post-traumatic stress disorders (PTSDs) can continue long after extreme events and disasters. The predictors of their persistence depend on factors such as the severity of the event, whether a life-threatening situation was experienced, history of psychiatric disorders, and exposure to additional traumatic events. Mental health impacts can be compounded for people or communities facing multiple stressors. For example, when combined with poverty, food insecurity, HIV/AIDS, a lack of water was found to contribute to both depression and anxiety among villagers in rural areas of Lesotho (Workman and Ureksoy 2017).

Although it is not possible to generalize about the type of reaction that any individual or group will have, it is clear that the characteristics and situation of the individual, their social context, and the event itself will shape emotional and psychological reactions and responses. In the case of Hurricane Katrina – a particularly severe event that struck the Gulf Coast of Louisiana in the United States in 2005 – recovery time for PTSD was influenced by a lack of access to basic necessities, forced relocation, and difficulty obtaining housing and employment (McLaughlin et al. 2011). Researchers found that people with higher family incomes took longer to recover from PTSD after Hurricane Katrina than those with low or low-to-average family incomes, a finding that was partly explained by whether or not existing worldviews were challenged (McLaughlin et al. 2011). The mental health impacts of climate change can intensify if community connections and social networks are disrupted or severed, which is common when people are displaced from their homes. Social exclusion has also been identified as a risk associated with lack of economic participation, lack of access to services, and a general lack of connection with community activities (Fritze et al. 2008).

Climate change, conflict, and migration

Many of the human security impacts of climate change that we have discussed so far will be felt most acutely by vulnerable populations that are already facing other forms of social and economic disadvantage. Climate change may also contribute to larger-scale security threats including the possibility of violent conflict or mass migration. Although projections of water wars and climate refugee crises are often featured in the media, the connections between climate change and these larger security threats are actually quite complex.

Conflict and peace

Concerns about climate-related conflict generally stem from the belief that higher temperatures and changes in rainfall will lead to disputes between groups or nations over increasingly scarce water and food supplies, and this, in turn, could escalate into violence. While it might seem logical to infer that resource scarcity will lead to violence, there is little evidence of a clear causal linkage between climate change and conflict (Adger et al. 2014). Whether reduced water availability (or any other environmental change) contributes to conflict depends on many factors, including institutions and governance systems. In the case of water resources, for example, these might include legal mechanisms for allocating water, informal agreements between users that allow them to manage shared resources, or protocols for arbitration of water disputes.

While climate change may not be a direct cause of conflict, it can under some circumstances be considered a security threat multiplier (Steinbruner et al. 2013). Climate change may influence factors that contribute to political and social stresses and insecurities, including worsening poverty, reduced access to or availability of natural resources, rising food prices, deteriorating terms of trade, changes in land tenure policies, or management decisions about water pricing and allocation. In the absence of effective institutions and governance, each of these factors can contribute to social unrest and political conflict. As Halvard Buhaug writes:

> Armed conflicts are more than anything else a symptom of political failure, both in terms of creating or allowing social conditions that provide fertile ground for widespread suffering and grievances (such as extreme poverty, inequality, oppression or corruption) and in terms of contributing to, or failing to prevent, social conflicts and protests escalating to the use of military violence. (Buhaug 2017)

Underlying social conditions such as high rates of poverty, inequality, and weak or failed governance are typically at the root of situations where climate-related stresses contribute to violence (Stern 2013; Adger et al. 2014). For example, the "Arab Spring" uprisings in 2011 have been traced to increased food prices following extreme weather in major wheat-growing regions in Canada, the United States, and Russia in 2011 (Johnstone and Mazo 2011). However, longer-standing social and political conditions, including high rates of unemployment and ineffective governing institutions, set the context for these protests.

Although climate change is rarely a direct cause of conflict, the presence of conflict and instability can exacerbate climate-related vulnerabilities and insecurities. Conflict damages infrastructure and life-support systems and undermines the capacity of institutions to prepare for and respond to climatic events (Barnett and Adger 2007). In countries located in the Horn of Africa, for example, the combination of drought and conflict, along with institutional failures, poverty, unemployment, and local environmental degradation, limits agricultural production and constrains other opportunities to achieve food security. As Sandstrom and Juhola (2017: 86) note, "food system failures in the Horn of Africa continue to be explained and labeled as droughts by the humanitarian community, with limited attempts to amend the humanitarian responses to address the complex [socioeconomic and political] causes of these food system failures."

In thinking about the potential for future conflicts in connection with climate change, it is important to recognize that changes in resource availability will be mediated through social institutions and the norms, rules, regulations, and rights that are in place to govern resources. Over the past three centuries, humans have initiated thousands of treaties to manage shared international water resources. This long track record of successful cooperation suggests that responses to climate change, rather than leading to conflict, may contribute to greater collaboration in the management of shared resources.

Migration and climate refugees

Climate change is often presented as an existential security threat for some communities and states, particularly small island states in the Pacific or Caribbean or communities living on the Bangladesh Delta. Population displacement, migration, and refugee crises are seen as likely outcomes of such threats. Despite such characterizations of climate change as a major driver of displacement and migration, there is relatively little evidence, to date, of large numbers of climate refugees (Black et al. 2011). Predictions of large-scale migration are mostly based on estimates of total populations that will be exposed to sea-level rise and other facets of climate change rather than on documentation of actual numbers of people that have been displaced by past or ongoing environmental changes (Gemenne 2011).

It is, in fact, difficult to even identify who qualifies as a climate change migrant or refugee. People make migration decisions for many reasons, including economic opportunities, family connections, or desire for political or religious freedom. Rather than a direct cause of migration, climate change may be better understood as one of many migration drivers, which also include a wide range of economic, political, social, and demographic factors. In a study of environmental change and migration in Niger, for example, Afifi (2011) found that the connections between climate stresses and environmental degradation were multifaceted. Interrelated problems of drought, deforestation, and soil degradation reduced incomes among farmers, herders, and fishermen which, in turn, contributed to economic migration. The migrants themselves, typically young men who left their families in search of work, identified economic factors such as poverty and unemployment as key reasons for migrating. However, nearly all indicated that environmental problems played a role in their decision to move.

While caution is needed when generalizing from past findings, it is nonetheless possible to envision numerous ways that climate change may contribute to a decision to migrate. Changes in availability of freshwater supplies, reductions in productivity of agricultural lands, disruptions to resource-related livelihoods, or damage to infrastructure and housing may all contribute to a future where there are many climate refugees.

Although we have been focusing on displacement or unplanned migration as a result of climate change, it is also important to keep in mind that migration is a key strategy for adaptation to environmental and climate stresses (Black et al. 2011; van der Land, Romankiewicz, and van der Geest 2018), as we will discuss in the next chapter. However, migration typically requires a significant outlay of financial resources, and actions such as selling property or livestock to cope with climate-related stresses can reduce a household's assets, such that individuals who wish to migrate may not be able to do so. As a result, the poorest of the poor may end up trapped in environmentally degraded areas that are highly exposed to climate shocks and stresses (Adger et al. 2014).

Summary

Climate change presents profound yet unevenly distributed threats to human security. For some vulnerable individuals and communities, climate change can be seen as a violation of basic human rights to food, water, and shelter, as well as a danger to health. For others, climate change may introduce new risks that threaten ways of life and emotional well-being. Climate change may also act as a threat multiplier which, under some circumstances, triggers larger-scale security threats. In the next chapter, we consider responses to climate change that fall under the broad umbrella of adaptation. Such responses are intended to help societies prepare for the impacts and reduce the risks of climate change.

Reflection questions

1. How and why is vulnerability to climate change connected to social and economic inequalities? Will reducing inequalities also reduce vulnerability?
2. Have you or anyone close to you experienced food insecurity, water insecurity, or any other type of insecurity following an extreme weather event?
3. Do you think that climate change may cause conflicts or wars in the future? In what ways can climate change responses promote cooperation and peace among nations?

Further reading

Gain, A. K., Giupponi, C., and Wada, Y., 2016. Measuring global water security towards sustainable development goals. *Environmental Research Letters* 11(12): 124015.

Klinenberg, E., 2015. *Heat Wave: A Social Autopsy of Disaster in Chicago*, 2nd edn. Chicago, IL: University of Chicago Press.

O'Brien, K. and Barnett, J., 2013. Global environmental change and human security. *Annual Review of Environment and Resources* 38(1): 373–91.

Ribot, J., 2014. Cause and response: Vulnerability and climate in the Anthropocene. *The Journal of Peasant Studies* 41(5): 667–705.

Wheeler, T. and Braun, J. von, 2013. Climate change impacts on global food security. *Science* 341(6145): 508–13.

9 ADAPTING TO A CHANGING CLIMATE

Adapting to change

We live in a world that is rapidly changing. We are constantly adjusting and adapting to new technologies, new job opportunities, new cultural experiences, and, increasingly, new climate conditions. Throughout this book we have emphasized that these changes are intertwined: development and globalization processes are contributing to climate change, and the impacts of climate change are shaping contexts and opportunities for development. Yet climate change is also different from most other changes that we experience because it introduces "severe, widespread and irreversible" global risks with far-reaching economic, social, political, cultural, and environmental consequences (IPCC 2014a: 17). These consequences will be unevenly distributed across regions, sectors, and social groups, yet they have implications for everyone. While some aspects of future climate change are unknown or uncertain, climate change involves many risks that can be reduced and managed with foresight and planning. One approach to reducing and managing the impacts of climate change is through adaptation.

What does it mean to adapt to a changing climate? How do we successfully adapt to changes while also reducing climate risks for future generations? In this chapter, we explore what adaptation involves and its implications for planning, development, and preparedness for future climate extremes. We discuss the concept of **climate resilience** and explain why successful adaptation to climate change involves not only adapting to climate change impacts, but also mitigating greenhouse gas emissions and taking actions to address the underlying drivers of risk and vulnerability. We also consider barriers and limits to successful adaptation and the potential for **maladaptation**. We conclude by recognizing that adaptation provides an important opening for activating individual and collective capacities for creating equitable, ethical, and sustainable futures.

What is adaptation?

Adaptation is generally understood as a change that is made in response to new conditions. It often connotes a new behavior, strategy, or intervention that is designed to accommodate or take advantage of a particular environmental or social change (see Box 9.1). According to the IPCC, adaptation to climate change is "the process of adjustment to actual or expected climate and its effects" (IPCC 2014d: 118). Adaptation can also refer to the result or outcome of these adjustments (Smit et al. 2000). As with other facets of climate change, interpretations of adaptation are closely linked to climate change discourses, as discussed in chapter 3.

Within the biophysical discourse, adaptation is often understood to be a technical, managerial, or behavioral intervention that minimizes exposure or reduces the severity of a particular climate impact (Biagini et al. 2014). Technical adaptations may include engineering measures such as raising bridges to accommodate higher water levels or installing irrigation systems in response to increasing agricultural demands for water. Technical adaptation measures can also entail taking advantage of opportunities associated with climate change, such as cultivating grapes to produce wine in regions where climates have warmed, or expanding cruise tourism into the Arctic region, where sea ice has diminished (Stewart et al. 2007; Ashenfelter and Storchmann 2016). Managerial and policy measures may include new regulations on how infrastructure is built or managed, establishment of early warning systems for storms, floods, or droughts, or changes in land-use zoning in order to influence where new construction is built and the types of structures that are allowed. Behavioral measures may include changes in the timing of agricultural practices or informational and educational campaigns that seek to raise awareness of dangers associated with extreme events such as heat waves.

While the biophysical discourse on adaptation tends to focus on reducing impacts or minimizing future exposure, the critical discourse draws attention to the root causes of vulnerability (Pelling 2011). These include economic and social processes and inequalities that put people in harm's way or make it difficult for some households or groups to respond. These processes may be linked to social and cultural norms, institutions, and power relations, as well as gender, racial, or ethnic discrimination (Jones and Boyd 2011). These and other forms of disadvantage contribute to uneven adaptive capacities among different individuals, households, groups, and communities. The most effective approaches to adaptation are understood to involve addressing root causes, including "on-the-ground processes of social differentiation, unequal access to resources, poverty, poor infrastructure, lack of representation and inadequate systems of social security, early warning and planning" (Ribot 2014: 5). The critical discourse also emphasizes that adaptation is not a politically neutral process; elite voices and vested interests often determine which adaptation strategies are selected, as well as where and when they are implemented (Eriksen, Nightingale, and Eakin 2015). In order to make adaptation processes and outcomes more equitable, the critical discourse emphasizes the need for inclusion and representation of vulnerable and disadvantaged communities in adaptation decision making (Chu, Anguelovski, and Carmin 2016). Recall that inclusion of a wide range of stakeholders is also a key element of the co-production of climate impact and risk assessment, as discussed in chapter 7.

The integrative discourse on adaptation recognizes that climate change is taking place within a dynamic social context that is influenced by numerous ongoing changes, including globalization, urbanization, and technological innovation (Wilbanks and Kates 2010). Together, these changes continuously present new challenges for individuals, communities, businesses, cities, regions, and nations, as well as international institutions. Responding to multiple processes of change, integrative approaches to adaptation emphasize the need for new pathways and possibilities for creating sustainable development trajectories (Wise et al. 2014; Fazey et al. 2016). Integrated strategies link adaptation, mitigation, and sustainable development, often by engaging a wide range of actors and appealing to values related to both the environment and quality of life (Shaw et al. 2014). This includes identifying outcomes of climate change adaptation or mitigation efforts that are valued within other discourses, such as social well-being (Bain et al. 2012). Such approaches recognize that individual and collective beliefs, values, and worldviews influence how problems are interpreted and addressed, and that adaptation to climate change may involve not only acknowledging multiple types of knowledge but also transforming our own individual and shared mindsets (O'Brien 2013).

Though the dismissive discourse does not consider climate change to be a serious concern, it nonetheless supports a range of approaches to adaptation. In some cases, adaptation measures, such as beach replenishment, early warning systems, or water conservation, may be taken, but the connection between these actions and climate change is downplayed or dismissed (Kuehne 2014). In other cases, the dismissive discourse may take a more fatalistic approach and assume that adaptation will occur autonomously, based on the idea that humans have always adapted to environmental change. In extreme cases, such as drought conditions, people may be encouraged to pray for rain. However, prayer as an adaptation strategy is not necessarily associated with the dismissive discourse. Studying climate change adaptation among cocoa farmers in Nigeria, Agbongiarhuoyi et al. (2013: 15) found prayer to be a prevalent response to climate change: "It is not surprising that cocoa farmers relied on prayers to God for rainfall and predicting climate change because most people call on God for solutions to problems at difficult times in Nigeria and West Africa." One commonality among adaptation responses within the dismissive discourse is that they do not recognize or address the ways that humans are contributing to climate change through energy use, agricultural practices, consumption patterns, and social and political decisions.

Although each discourse interprets adaptation differently, it is clear that adaptation to climate change is a social process that involves bringing people together to make decisions about actions and strategies to meet changing conditions. This includes decisions on whether, where, when, and how to invest resources, how to minimize loss and damages, and how to deal with people who are displaced from their homes or searching for new livelihoods (Bhattarai, Beilin, and Ford 2015; Chapin III et al. 2016). How we think about adaptation is ultimately linked to how we define the problem, what solutions we think are possible, and which responses we prioritize.

Adaptation in practice

Adaptation to a changing climate is already underway. As we have discussed in previous chapters, global average temperatures have increased by nearly 1°C since the late nineteenth century, with much of this warming occurring in recent decades. Sea levels are rising, rainfall patterns are changing, and the climate is becoming more variable. Households, communities, cities, and sectors throughout the world are preparing for more extreme weather events and coastal flooding and are making provisions for added stresses on water supplies, public health services, and infrastructure. Climate change adaptation is also being incorporated into development policies and practices, often in technical or instrumental ways or through capacity-building activities (Inderberg et al. 2014). In considering future strategies and options for adaptation, it is important to keep in mind the enormous differences between the low-end scenarios and high-end scenarios of climate change, not only in terms of future temperatures but also with respect to tipping points that produce abrupt system-wide changes (Lenton 2013). The differences between a 1.5°C versus a 4°C increase will be profound in terms of the types of adaptations that are needed, the costs associated with them, and losses and damages that are incurred.

Adaptation strategies

As adaptation takes a more visible role on planning agendas in countries, cities, communities, and businesses, increasing attention is paid to the practicalities of selecting, evaluating, and implementing various strategies. Depending on timing and intention, adaptation strategies may be reactive, passive, or anticipatory. Reactive adaptations include actions that are deliberately taken in response to the experienced impacts of climate shocks and stresses. For example, in a study of adaptation to flood risks in seven countries, Kreibich et al. (2017) identified numerous adaptations taken in response to previous flood events. These reactive responses, which ranged

Box 9.1 Adaptation, history, and evolution

While adaptation to climate change is often described as a new or unprecedented challenge for humanity, adaptation has been foundational to the evolution and development of human societies. Throughout history, humans have adapted to changing environmental conditions by pursuing different livelihoods, developing new technologies, establishing new institutions, or migrating to new areas. Such behavioral and cultural adaptations have allowed humans to survive and thrive during periods of climate variability and change. Failure to adapt to changing conditions has, in some cases, contributed to crisis, decline, or even collapse of societies (Butzer and Endfield 2012). Some examples where local or regional climate changes are thought to have played a significant role in the decline of societies include the Classic Maya of Mexico and Central America and the Viking settlements in Greenland (see Orlove 2005). Yet it is seldom possible to attribute the downfall or collapse of societies to environmental or climatic changes alone. Institutional and political factors, social and cultural contexts, and other contemporaneous events have played a vital role in the decline of societies or cultures. Drawing lessons from historical cases, Butzer and Endfield (2012) identify factors such as collaboration, openness to change, and willingness to challenge assumptions as key factors in successful societal adaptations.

Humans are not the only ones that have had to adapt to changing conditions. Species and ecosystems are also continually adapting in response to both natural and human-induced environmental stresses. For species, adaptation may involve shifting geographical distributions and the timing of growth and reproduction, or it may demand evolutionary responses, such as genetic adaptations. While evolutionary adaptations can influence community dynamics in response to changing climatic conditions, adaptation opportunities for other species are often constrained by human activities (Hoffmann and Sgrò 2011). Expansion of urban settlements, deforestation, habitat fragmentation, damming of rivers, contamination of air, soil, and water, and other stresses affect adaptation options for both species and ecosystems. Although some efforts are underway to protect particular species by, for example, constructing biodiversity corridors to allow migration to new areas, many species will have difficulty adapting to new climate conditions.

Do you think that society's past adaptations to environmental stresses can help us to prepare for climate change? With respect to other species, do you think humans should try to help them to adapt to the impacts of climate change? Why or why not?

from the securing of household oil tanks in Germany to expansion of early warning systems in Spain to relocation of households from flood-prone regions in Vietnam, were found to reduce vulnerability to subsequent flood events. At the individual and household level, reactive adaptations often tend to be autonomous, carried out spontaneously in response to an experienced impact. For example, in regions where extreme heat is making it dangerous to work outdoors, some individuals are autonomously adapting their work schedules to being outside earlier in the morning or later in the evening. These adaptations, which are intended to reduce heat exposure, have implications for everything from energy usage to transportation to family mealtimes.

Passive adaptations include actions and adjustments taken as a result of changing environmental conditions but not purposefully intended to respond to climate change. Tripathi and Mishra's (2017) study of adaptation among farmers in Uttar Pradesh, India, for example, showed that most of them were changing their farming practices in ways that enhanced capacity to cope with a changing climate, but that these actions were not perceived as a response to climate change. Passive adaptations may also be made by those who are skeptical about the causes of climate change. For example, a study by Kuehne (2014) found that, despite their disbelief in climate change, Australian farmers were taking numerous adaptation actions. These passive adaptations were directed at meeting immediate challenges such as water scarcity, low commodity prices, and the rising costs of inputs such as fertilizer, but they were not interpreted as connected to climate change.

Anticipatory adaptations include actions and strategies that are planned in advance (Smit and Wandel 2006). They may entail proactive policies that are adopted prior to anticipated future climate conditions, such as higher sea levels, heavier rainfall events, or more severe heat waves. Such adaptations, which are intended to avoid negative impacts or reduce future vulnerabilities, require considerable foresight and planning. For example, in anticipation of warmer temperatures in the future, apple farmers in Japan may decide to plant different varieties or different types of fruit, such as peaches (Fujisawa et al. 2015). Because apple trees grow for a number of years before they bear fruit, apple farmers cannot shift varieties from year to year. Planning is necessary for most anticipatory adaptations, particularly for large-scale infrastructure such as public transportation systems that are expected to last for many decades. For example, transportation and water-system planners in cities such as Copenhagen who are developing new subway and sewerage systems are proactively accounting for an increased potential for flooding and sea-level rise in the design of these systems (City of Copenhagen 2011).

The distinction between reactive, passive, and anticipatory adaptation is often closely linked to the question of *who* is doing the adapting and *what* is being adapted (Smit et al. 2000). Many actors are involved in adaptation decisions. Farmers, fishermen, and others working or making a living in sectors that are highly exposed to climate change may take reactive, passive, or anticipatory steps to adapt, whether by altering the timing of planting or types of fertilizers, diversifying crops, changing the areas in which they fish, targeting different fish species, or taking steps to secure and protect their livestock. The international business community is also paying more and more attention to adaptation, particularly in cases where climate change threatens their products or services. For example, companies such as Levi Strauss have been proactively working in the cultivation of cotton and the manufacturing process for blue jeans to reduce their environmental impacts, particularly in relation to water usage (Joule 2011). Chocolate companies recognize that their supply chains are threatened by reduced cacao yields, as many cacao-growing regions are susceptible to drought and to the combined effects of hot and dry conditions. In anticipation of these threats to production, they are investing in genetic breeding to develop drought- and disease-resistant cacao varieties (Farrell et al. 2018).

At the community, city, and regional levels, adaptation actions are often implemented through government agencies or institutions that are already responsible for infrastructure systems, environmental protection, disaster-risk management, public health, and community development (Mullan et al. 2015). A small group of individuals working as planners or resource managers within different government agencies are often responsible for preparing and implementing adaptation plans. They have to make practical decisions within budgetary constraints and may face resistance from homeowners, builders, preservationists, tourism developers, and others with competing interests. Their decisions also need to take into account various laws, rules, and regulations, and they may also want to include citizens and other stakeholders in adaptation decision making. Policy processes that are perceived to be top-down rather than participatory can sometimes lead to mistrust, opposition, and defiance. In a study of stakeholder perspectives on adaptation in coastal zones of the United Kingdom, Few et al. (2007) found that anticipatory adaptation efforts that involved long-term planning and financial commitments magnified the challenges of involving stakeholders. These challenges arose due both to the uncertain nature of future climate change and to a lack of perceived short-term benefits of adaptation investments.

Community-based adaptation (CBA) offers an alternative approach to incorporating local needs in the selection of appropriate responses (Chapin III et al. 2016). In contrast to top-down models of citizen involvement, CBA approaches deliberately engage people in a collaborative decision-making process that attempts to better align adaptation planning with the needs, interests, local knowledge, and cultural context of residents (Ensor and Berger 2009). While CBA approaches have the potential to make adaptation decisions more inclusive and democratic, they may also have unintended consequences. In their review of efforts to promote CBA in Inuit communities in the Canadian Arctic, Ford et al. (2016) found that some approaches may have undesirable outcomes, such as creating a perception that adaptation is solely a local issue that does not require regional- or national-level action. As we discuss in the next section, undesirable outcomes of adaptation are frequently referred to as maladaptation.

Maladaptation

In considering the viability of different adaptation practices and actions, it is important to recognize that adaptation may have unintended consequences or contribute to new problems. Barnett and O'Neill (2010: 211) define maladaptation as "action taken ostensibly to avoid or reduce vulnerability to climate change that impacts adversely on or increases the vulnerability of other systems, sectors, or social groups." For example, one community's adaptation to water scarcity might involve digging deeper wells or diverting water from a river, leaving other communities more vulnerable to water shortages. Maladaptation may also contribute to increased exposure to climate risks. Construction of flood walls as an adaptation to sea-level rise may result in additional development and new settlement of coastal zones, thereby adding to the amount of property that is potentially in harm's way. Another example is the purchase of flood insurance, which has long been used as a means of distributing risks and sharing the burdens of extreme weather events and is increasingly viewed as a potential strategy for climate change adaptation. However, flood insurance can also have unintended consequences, including promotion of property development in highly exposed areas (Thomas and Leichenko 2011).

Maladaptation to climate change also results from actions that increase emissions of greenhouse gases and contribute to additional climate change. An often-cited example of maladaptation is the increasing use of air conditioners to cool indoor spaces as temperatures rise. Electric air conditioners remove heat and humidity from homes, office buildings, cars, and shopping centers, and they have played a key role in the development of large urban settlements in hot climates. Although air conditioning is expected to be a key adaptation to climate change, its growing usage also has the potential to increase greenhouse gas emissions if electricity comes from fossil fuel energy, thereby contributing to additional climate change (International Energy Agency 2018) (see Figure 9.1). Some coolants used in air conditioning, such as hydrofluorocarbons, contribute to both rising greenhouse gas emissions and ozone depletion (Reese 2018).

Maladaptation also describes actions that burden the most vulnerable or reinforce existing inequalities and environmental injustices. For example, one

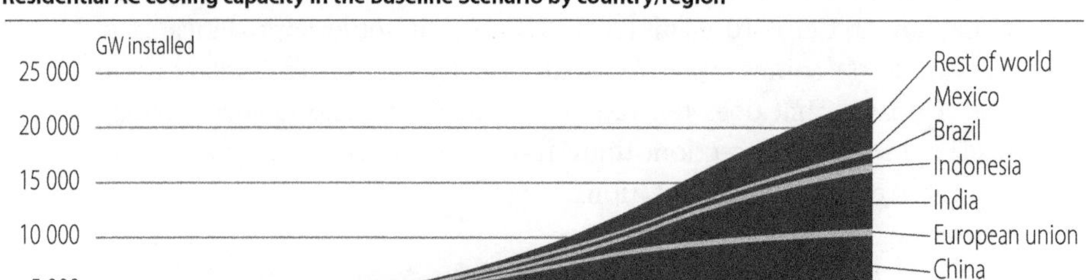

Figure 9.1 Air conditioning rates.
Source: International Energy Agency 2018

agricultural adaptation to the spread of disease and plant pathogens under warming climates might involve greater application of pesticides on crops. Higher pesticide usage not only adds to chemical loads in the environment but may also have unintended health effects for agricultural workers, who already face significant health burdens from chemical inputs (Wolfe et al. 2011). In order to ensure that adaptation measures do not increase the vulnerability of poor or other disadvantaged populations, Eriksen et al. (2011) identify several principles of sustainable adaptation. These include (1) recognizing the context that contributes to vulnerability; (2) acknowledging how different values and interests influence adaptation outcomes; (3) integrating local knowledge; and (4) considering potential feedbacks across scales. Sustainable adaptation measures typically combine vulnerability reduction and poverty alleviation through, for example, construction of affordable housing in areas that are safe from climate hazards, improvement of access to credit, and strengthening of traditional systems of communal support (Leichenko and Silva 2014). By paying deliberate attention to the needs of poor and disadvantaged populations, such measures can help to minimize the possibility of maladaptation.

Cost–benefit and risk-management approaches

Adaptation decisions require information about the costs of different strategies and the potential risks that might be incurred, whether or not adaptation actions are taken. A method known as cost–benefit analysis is often applied to adaptation decisions to assess whether particular actions or policies make sense from an economic standpoint. In the case of agriculture in a drought-prone region, for example, an irrigation system might be proposed as an adaptation to help farmers cope with increased rainfall variability. An assessment of costs and benefits of this strategy would entail consideration of the costs of installation and operation of the irrigation system in relation to benefits, including increased crop yields or addition of a second cropping season. Estimating the net effects of adaptation entails comparison of the costs of crop losses that would be incurred without irrigation versus its costs and expected benefits (Leichenko et al. 2011). With respect to sea-level rise, a cost–benefit study may compare the costs of construction of flood walls or levees with the benefits they offer in terms of protection of property, road infrastructure, port facilities, and other assets.

Cost–benefit approaches are widely used by governments and international development agencies to select which adaptation options are likely to be most beneficial (see Box 9.2). These approaches are also applied to policies and actions that are intended to prevent climate change, such as the implementation of new regulations on greenhouse emissions from power plants. However, many non-monetary impacts of climate change, such as loss of culturally valued assets and knowledge and a community's sense of place, are difficult to capture in these assessments. Moreover, some types of impacts, such as species extinction or loss of a state or country to sea-level rise, are nearly impossible to quantify because they include so many intangible aspects.

As an alternative to cost–benefit assessment, adaptation to climate change can be viewed through a lens of risk management (Jones and Preston 2011; Solecki, Pelling, and Garschagen 2017). Risk management is something that individuals do on a regular basis in their everyday lives. For example, many people purchase fire, auto, and health insurance to protect themselves from exposure to financial loss in the event of an accident or illness. Similarly, business firms purchase insurance in order to reduce their exposure to serious financial losses in the event of a fire, power outages, cyberattacks, or other threats. Understanding of risks is also inherent in managing complex systems such as global production chains for manufacturing, electricity supplies, and cellular communication networks. A risk-based approach to adaptation assumes the impacts of climate change can be managed in ways that are similar to how other types of weather risks are managed, such as extreme storms, floods, and heat waves (Jones 2001; Yohe and Leichenko 2010). Because extreme weather already affects every society, use of risk-based approaches also offers an opportunity to coordinate efforts to promote reduction of all types of weather-related disaster risks with climate change adaptation (IPCC 2012; Serrao-Neumann et al. 2015).

Within a risk-management framework, risks are described in terms of the probability of a particular weather event such as a heat wave or storm multiplied by the consequences of that event:

Risk = (probability of an event) * (consequences of the event)

The probability of an event is a statistical estimate of the odds or likelihood of that event occurring. The consequences describe the expected impacts of the event. Under climate change, both probabilities and consequences of extreme events may change, with implications for adaptation policies and practices.

Risk-based approaches explicitly recognize the need to develop **flexible adaptation pathways** for responding to a wide range of climatic shocks and stresses (Yohe and Leichenko 2010) (see Figure 9.2). These approaches suggest that planning decisions should be made with the expectation that formerly rare climate events may become more common in the future (Milly et al. 2008). For example, many types of infrastructure, such as municipal water systems, are designed to withstand the effects of a rainfall event that is expected to occur once every one hundred years. This is commonly referred to as a 100-year storm because, in any given year, such an event has a 1%

Box 9.2 Cost–benefit analysis and discount rates

One major challenge in any type of cost–benefit assessment is how to compare the costs of investments made today with the value of benefits that will accrue in the future. A discount rate (also known as an interest rate) is typically applied in these calculations to determine the present-day monetary value of future benefits or expenditures (Leichenko et al. 2011). The choice of which discount rate to use and the time period over which future benefits are discounted heavily influences the results of the analysis.

The selection of different discount rates and time horizons are not only a methodological question. They also introduce moral and ethical debates about how we value the environment and the future. A very high discount rate, for example, implies that money spent today is far more valuable than money spent in the future. The argument for high discount rates is based on the expectation that society will be richer in future and will be more easily able to afford adaptations (Stern 2013). By contrast, a very low or zero discount rate implies that money spent today would have a similar value in the future. When the present and future are valued more equally, there is a stronger case for taking adaptation (and mitigation) actions in the present. The choice of discount rates and time horizons leads to a wide range of estimates of the costs of adaptation to climate change (Chambwera et al. 2014).

Do you think that costs associated with climate change impacts that happen in the future should be discounted compared to impacts that happen today? Do you think society should make investments today to avoid costs associated with future impacts?

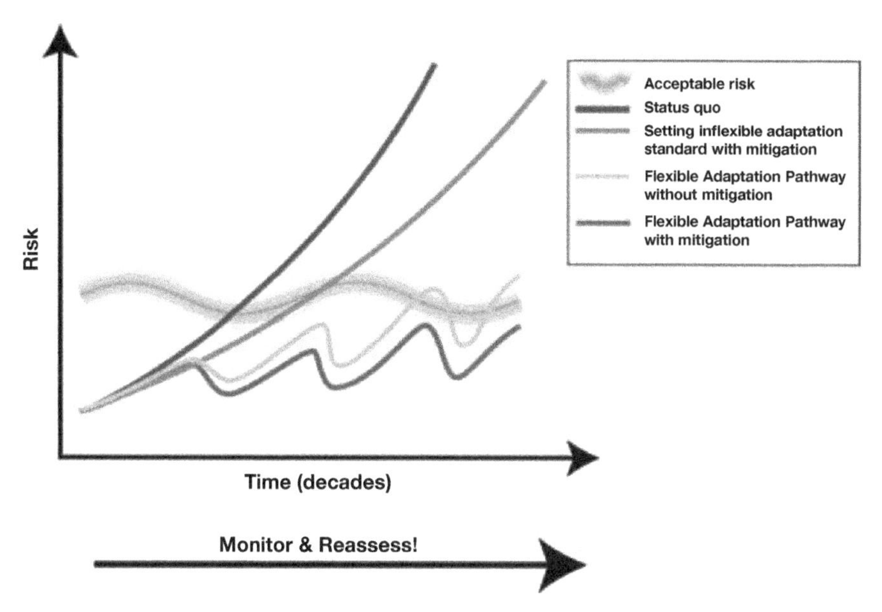

Figure 9.2 Flexible adaptation pathways.
Source: Yohe and Leichenko 2010

chance of occurring (i.e., a 1 in 100 probability). Because water systems are often designed to withstand 100-year events, the consequences of such events are expected to be fairly minimal. By contrast, the consequences of a more extreme rainfall event, such as a 500-year event, can be quite devastating. Such an event may cause inundation and contamination of water-purification ponds and damage to intake water pipes, thereby compromising water supplies. As a result of climate change, the probabilities of extreme events are increasing, such that formerly unlikely extreme events, such as a 500-year flood, are expected to occur more frequently. Although identifying acceptable levels of risk for water supplies and other vital systems still involves normative judgments, taking into account the changing probabilities of extreme events provides valuable information for future planning.

An important limitation of both cost–benefit analysis and risk-based approaches to adaptation decisions relates to the complexities of predicting future development trends and climate change. In addition to the uncertainties associated with future emissions scenarios, it may not be possible to assign probabilities to many facets of climate change because of non-linearities in the climate system, interactions with other systems, and thresholds or tipping points that may create irreversible changes and other unknowns and surprises (Stern 2013). It is particularly difficult to capture costs and risks that are associated with emerging threats that could have catastrophic outcomes, such as depletion of food supplies, the spread of new vector-borne diseases, or cascading infrastructure failures (Oppenheimer et al. 2014). These approaches may also downplay the role of economic, political, and cultural contexts in shaping the consequences of extreme events, and they offer little insight into how costs and risks are distributed across society (Wisner et al. 2003; Hulme 2014).

Building climate change resilience

The notion of building resilience to climate change offers another approach to adaptation planning. Resilience entails identifying and fostering characteristics that allow individuals, communities, cities, and systems to effectively respond, recover, and return to "normal" functioning following a climate-related event, and to respond better to future impacts (Leichenko 2011). Building climate change resilience enhances the capacity to absorb and bounce back or recover from climate change shocks and stresses. The concept of resilience can be applied to many facets of adaptation. For example, psychologists may explore resilience of individuals to traumas related to extreme events and disaster; engineers may assess the resilience of infrastructure systems in terms of capacity to function under extreme weather conditions or gradual, long-term changes; economists may measure resilience of a region's economy based on how quickly it recovers from a major hurricane or flood; and ecologists may consider the resilience of an ecosystem in terms of its ability to withstand climatic perturbations or regenerate in their aftermath.

As with other approaches to adaptation, understandings of resilience are influenced by particular discourses. Resilience strategies may focus on technical measures such as promoting the use of green infrastructure, social changes such as fostering new community-based networks, or changes in perspectives such as encouraging mindsets that embrace flexibility, learning, and openness to change. While resilience can sometimes be interpreted as a justification for maintaining or perpetuating the status quo, it can also include the idea of "bouncing forward." This interpretation places emphasis on building the capacity to anticipate climate shocks and respond in ways that reduce future vulnerabilities and contribute to sustainability (Manyena et al. 2011). Whether talking about resilient communities, cities, or systems, climate resilience implies being able not only to live with and manage change but to recognize the conditions and contexts where systems and structures have to change in order to reduce future risks and contribute to wider goals of sustainable development (Denton et al. 2014).

Efforts to build resilience are often centered on the community level, which is a focus for many planning initiatives (Cutter et al. 2014). Resilience-oriented strategies start by identifying characteristics of communities that make them more robust in relation to climatic shocks and stresses. These factors often have to do with effective institutions, human and social capital, economic diversity, and governance (Leichenko 2011). For urban communities, building resilience may entail preparation for climate extremes through the expansion of early warning systems and development of post-event rescue and recovery plans. For communities that are dependent on agriculture, building resilience might involve a combination of diversification of crops grown and identification of new livelihood options. While the particular strategies to build resilience will vary by community, these approaches share the common goal of ensuring that communities are able to respond to and recover from the consequences of climate change.

The concept of climate resilience is also being applied to urban infrastructure systems. Extreme events that we have discussed in earlier chapters, such as Hurricanes Sandy in New York City and Maria in Puerto Rico, flooding in Bangkok and Chennai, and drought in Cape Town, have revealed significant vulnerabilities in energy, transport, and water supply systems. Building climate change resilience into energy grids and other infrastructure systems often entails engineering measures such as enhancing capacity of these systems to withstand high winds, extreme heat, and heavy rainfall. These measures may also include efforts to avoid outages or disruptions by, for example, trimming trees around overhead power lines or mandating that commercial and industrial customers reduce energy usage during heat waves or other periods of high demand. In Puerto Rico, efforts are underway to build a more resilient electricity system based on distributed renewable energy and micro-grids (Gallucci 2018). For industrial firms, installation of off-grid energy generation capacities, such as wind turbines or solar panels, can offer protection from system outages while also reducing emissions. Such strategies also have the co-benefit of reducing overall energy costs (Smith and Sweet 2013).

A focus on resilience may also help to guard against some forms of maladaptation by emphasizing strategies that promote the capacity both to recover from climate shocks and to foster reduction of risks through mitigation of climate change. For example, putting a green roof on a building can be thought of as both an adaptation and a mitigation strategy. A green roof entails planting vegetation on the top of a building, which can keep it cooler during very hot weather and also reduce the demand for energy for air conditioning. Green roofs may also provide co-benefits, such as aiding stormwater management, increasing urban biodiversity, sequestering carbon, reducing air and noise pollution, and providing a space for urban agriculture (Demuzere et al. 2014).

While resilience-building is arguably becoming a dominant strategy for adaptation among communities and cities worldwide, the issue of equity is a lingering concern, including the possibility that some resilience-building strategies may negatively affect poor communities or may prioritize the needs of elite groups at the expense of others (Anguelovski et al. 2016). By failing to consider whose livelihoods and environments are being protected, resilience approaches may inadvertently reinforce unequal power relationships and inequalities (Tschakert and Dietrich 2010; Cote and Nightingale 2012). In order to ensure that resilience efforts do not contribute to maladaptation for poor communities, cities such as Boston are explicitly focusing on ensuring that socially vulnerable populations and neighborhoods are not disproportionately harmed by climate extremes (City of Boston 2018). However, building equity into resilience is not simply about targeting efforts towards poor communities but also requires attention to the decision-making process. Adaptation efforts that are designed using principles of CBA, sustainable adaptation, and co-production can help to ensure that interventions take into account the needs and wishes of community members (Sarzynski 2015). In the case of communities that already face significant burdens due to poverty and exposure to toxic wastes, for example, steps to enhance resilience may need to go beyond enhancement of response or recovery capacities to address larger questions of environmental and social justice. Depending upon local needs and wishes, measures to ensure resilience of such communities may range from removal or cleaning of toxic waste sites to investment in human capital through improved education and job training, to grants for local renewable energy generation projects. As we discuss in the next chapter, inclusive and equitable responses to climate change that incorporate both adaptation and migration are fundamental for transformations to sustainability.

Barriers and limits to adaptation

It is widely assumed that adaptation is a reasonable response to climate change, particularly if greenhouse gas emissions cannot be significantly reduced. While much can be done in theory to adapt to climate change, in practice there are both **barriers to adaptation** and **limits to adaptation**. Barriers include political, economic, technical, informational, and cultural factors that can hinder adaptation planning and implementation (Leichenko, McDermott, and Bezborodko 2015). Limits are generally understood as absolute or fixed obstacles that render adaptation ineffective (Islam et al. 2014). Although barriers are expected to be surmountable, in practice they may serve as limits to adaptation if resources and capacities cannot be mobilized to overcome them.

Barriers to adaptation often stem from mindsets and cultural norms. For example, institutional cultures that favor technological measures, such as construction of flood walls to protect an area from storm surges, may impede consideration of other types of responses, such as changes in land use

(Jeffers 2013). While technological measures may offer short-term protection for some areas, in the long term they may not be as effective as relocation out of flood-prone areas. Barriers can also arise based on presumptions by decision makers that private property rights and land uses must be protected (Few, Brown, and Tompkins 2007). For example, a proactive adaptation to climate change in some coastal areas may involve restricting new housing construction in low-lying locations. However, local governments may be reluctant to propose new land-use regulations that may conflict with private investments or affect the value of existing homes. Social and cultural norms (such as notions of personal freedom) and lack of knowledge and information about climate change and can also present barriers to adaptation action (Jones and Boyd 2011; Lata and Nunn 2012). While some barriers to adaptation can be overcome, others may be insurmountable for particular communities or sectors. For example, as weather-related losses mount, insurance companies may begin to take a closer look at climate change projections and consider withdrawing insurance coverage in some high-risk areas, such as flood-prone communities in Queensland, Australia (McAneney et al. 2016).

Questions of how to pay for adaptation measures and who should be responsible for these costs may also present barriers to adaptation. Adapting to climate shocks and stresses is likely to require considerable resources, potentially beyond what individuals, communities, and businesses can afford. As we discussed in earlier chapters, many island nations and coastal communities are already threatened by rising sea levels and storm surges. Here, questions of how to finance adaptation are an immediate concern. Regardless of whether residents and businesses in these communities wish to remain or relocate, issues of whether and how to financially support adaptation will need to be addressed (see Box 9.3).

Paying for adaptation presents an especially high barrier for poor populations. Strategies that are commonly adopted in wealthier countries, such as the purchase of insurance, may be largely unaffordable for the most vulnerable, especially in less industrialized countries and regions (Kunreuther and Lyster 2017). Although there has been considerable discussion of the role of microinsurance and other financial mechanisms in adaptation, insurance coverage for disaster losses is highly variable worldwide, and it is considerably lower in Asia, Africa, and Latin America than in Europe, North America, and Australia (see Table 9.1). Several international programs have been established to help fund adaptation projects in developing countries, such as the Global Environmental Facility (GEF) and the Green Climate Fund (GCF), but these institutions often support adjustments that are consistent with development as usual, rather than measures that address the root causes of risk and vulnerability (Bassett and Fogelman 2013; Eriksen et al. 2015).

There are likely to be many instances where adaptation will be an insufficient response to the risks associated with climate change (Adger et al. 2009;

Table 9.1 Uninsured disaster losses by region in 2017

Region	Uninsured losses in %
Asia	92
South America	89
Africa	87
Europe	65
North America incl. Central America and Caribbean	53
Oceania	45

Source: Compiled with data from Munich Re NatCatSERVICE 2018

Dow et al. 2013; Barnett et al. 2015). Sea-level rise in particular is expected to pose long-term challenges for adaptation, as the processes and consequences are likely to last for millennia (Levermann et al. 2013). There may also be physiological limits to adaptation for both humans and non-humans. There are, for examples, physiological limits to tolerance of heat stress. While humans may be able to devise protections against unprecedented heat stress, such protections may not be affordable, satisfying, or effective for most of humanity (Sherwood and Huber 2010). For species that are native to particular biomes, such as Arctic ecosystems, coral reefs, and tropical cloud forests, the capacity to adapt to new climate conditions may be quite limited (see Box 9.1).

Discussions of the limits to adaptation often focus on ecological and physical conditions representing critical thresholds, or states beyond which existing activities, land uses, and ecological systems will be unable to adapt without radical transformations (Moser and Ekstrom 2010). Although some limits are technical, ecological, economic, and physiological, others may be subjectively defined by what a group values, such as the experience of snow in winter (O'Brien 2010). Limits also depend on the goals of adaptation, which may be linked to diverse cultural values, risk perception, and ethics (Adger et al. 2009). Consequently, limits to adaptation may be better understood as socially and culturally contingent rather than absolute (Dow et al. 2013).

Recognizing the limits to adaptation also means acknowledging that there will be significant losses and damages associated with climate change. Losses and damages refer to the economic and non-economic consequences of climate change that cannot be avoided through either adaptation or mitigation (Warner and van der Geest 2013). These may include loss of life, property, and cultures. They also include losses of ecosystems, species, habitat, and ecosystem services. The question of whether and how to compensate affected countries for loss and damages from climate change has become an important part of international climate change negotiations.

Box 9.3 Should we save Tangier Island?

Tangier Island, Virginia, is one location that is considered to be on the front line when it comes to experiencing the impacts of climate change. Homes, businesses, and infrastructure on this low-lying island located in the Chesapeake Bay are already experiencing significant damage as a result of sea-level rise and other factors (Gertner 2016). The immediate costs of adaptations such as pumping in new sand, adding breakwaters, and planting new vegetation are estimated to be on the order of US$20–30 million (Schulte, Dridge, and Hudgins 2015). While some residents want to leave the island, others would prefer to remain. This is not an easy decision for residents of Tangier Island, and it also raises broader questions about adaptation to climate change and the limits to adaptation.

If you lived on an island threatened by sea-level rise, would you prefer to stay or leave? If you wanted to stay, do you think others should help pay the costs of adaptation measures? If so, why? If you would rather leave, do you think others should help pay for you and your family or community to relocate? Now imagine that you live on the mainland. Do you think residents of the mainland should help pay the costs of helping island residents to either adapt or relocate? Why or why not?

Many of the countries that will experience the greatest losses from climate change – particularly small island nations – have argued that there is a need for compensation for loss and damages from nations that are most responsible for contributing to climate change.

One outcome of these negotiations is the Warsaw International Mechanism for Loss and Damage, which was established in 2013 as part of the United Nations Framework Convention on Climate Change (UNFCCC). The mechanism is intended as a way to "address loss and damage associated with climate change in developing countries that are particularly vulnerable to the adverse effects of climate change in a comprehensive, integrated and coherent manner" (UNFCCC 2018). While efforts such as the Warsaw Mechanism are a useful starting point for addressing losses associated with climate change, there are continuing debates and disagreements over how losses and damages should be counted, what should be included in these calculations, and who should pay for them (Gewirtzman et al. 2018).

Summary

Adaptation will be necessary for the foreseeable future. Even if greenhouse gas levels were to stabilize today, warming of the Earth and melting of glaciers and ice caps will continue for many decades. Yet, as we have shown in this chapter, adaptation to climate change does not have to be merely a reactive response to climate change impacts. Through social learning and collaboration, society has the potential to respond to global risks in new ways. While history suggests that human potential for adaptation is high, the question is whether and how this potential will be realized. Ultimately, adaptation to climate change may offer an opportunity for humans to redefine their relationships with the environment, with each other, and with the future. In the next chapter, we consider the future and explore the types of transformations that may be needed to meet the challenge of climate change in an ethical, equitable, and sustainable manner.

Reflection questions

1 Who do you think should be responsible for paying for adaptation? Those who are affected by the impacts of climate change or those who are responsible for increasing emissions?
2 Does building resilience seem like a good strategy for preparing for the impacts of climate change? Are there limits or drawbacks to resilience?
3 Think about an activity or place that you value which is threatened by climate change. What type of loss would you experience in the event that adaptation fails to preserve or protect this activity or asset?

Further reading

Adger, W. N., Dessai, S., Goulden, M., et al., 2009. Are there social limits to adaptation to climate change? *Climatic Change* 93(3–4): 335–54.

Barnett, J. and O'Neill, S., 2010. Maladaptation. *Global Environmental Change* 20(2): 211–13.

Cutter, S. L., Ash, K. D., and Emrich, C. T., 2014. The geographies of community disaster resilience. *Global Environmental Change* 29: 65–77.

Eriksen, S. H., Nightingale, A. J., and Eakin, H., 2015. Reframing adaptation: The political nature of climate change adaptation. *Global Environmental Change* 35: 523–33.

Pelling, M., 2011. *Adaptation to Climate Change: From Resilience to Transformation*. New York: Routledge.

10 TRANSFORMING THE FUTURE

Imagining the future

Imagine the year is 2080, and you are looking outside the window. You are likely to see a world that is dramatically different from today. You might be living in a place where children can no longer play outdoors during the summer because it is too hot, where water has to be imported from distant places, and where sea levels have risen substantially, submerging coastal infrastructure, property, and historical sites and displacing people from their homes. Or you might be looking at a place where solar mobility units are transporting thousands of people across the city every minute, where food gardens form a ribbon along what used to be highways, where birds and insects are abundant, and where greenhouse gas emissions are decreasing, or at least have stabilized. As we look ahead, it is useful to pause and consider what type of futures *you* imagine are possible – or impossible. What type of future do you want?

There is no doubt that we will be living with climate change and its consequences for the foreseeable future. We also know that, without significant actions, climate change will have unprecedented consequences for natural and social systems. It is worth repeating that the Intergovernmental Panel on Climate Change (IPCC 2014a: 17) concludes with high confidence that, "Without additional mitigation efforts beyond those in place today, and even with adaptation, warming by the end of the twenty-first century will lead to high to very high risk of severe, widespread, and irreversible impacts globally." Yet a continually warming future is *not* inevitable. The identified risks of climate change are long-lasting and unequally distributed, but many of them are also avoidable. As we have emphasized throughout this book, the differences between the low-end scenarios and high-end scenarios of climate change are profound. The type of future that we experience will, to a large extent, depend on human choices, decisions, and actions.

In this chapter, we discuss transformative responses to climate change. We first define transformation and discuss why transformations to sustainability are needed. We then present an integrative lens on transformation that emphasizes connections among the practical, political, and personal spheres. We also explore various entry points for transformations that promote a sustainable, equitable, and thriving world. We show that these shifts are not only possible but already underway. We conclude by exploring the role of individuals and groups in activating change, emphasizing that each of us has the opportunity to contribute to a more sustainable future.

The future we want

As more people recognize that the future will be influenced by individual and collective decisions and actions taken today, there are increasing calls for transformations that are not only sustainable but also ethical and equitable. "The Future We Want" is a UN resolution and call to action that emerged from the 2012 United Nations Conference on Sustainable Development (referred to as the Rio+20 Summit). The resolution highlights the connections between climate change, poverty eradication, and sustainability, emphasizing that "people are at the centre of sustainable development and, in this regard, we strive for a world that is just, equitable and inclusive" (United Nations General Assembly 2012). Realizing such a world means paying attention not just to the goals and outcomes of transformations, but to the process itself.

History tells us that societies can and often do transform to meet new challenges. In many cases, these transformations arose from a change in perspective, from seeing relationships from a different point of view, or by challenging outdated assumptions about human behavior or how societies are organized (Hochschild 2006; Butzer 2012). For example, the mechanized printing press, invented in the fifteenth century in Germany, spread quickly in Europe and had widespread political consequences, leading to the emergence of an educated middle class as well as greater cultural self-awareness. More recent historical examples of transformation, such as the abolition of slavery in the United States or the granting of voting rights to women in most countries, remind us that transformations often involve struggles and are typically met with resistance related to power, politics, and interests. Many social movements champion universal values and entail a recognition of rights and responsibilities, including holding governments, companies, and people accountable for their actions. Most successful movements involve both individual and collective action and a strong sense of political agency. In the context of climate change, political agency can be thought of as the capacity to positively influence the collective future through transformative change (O'Brien 2015).

Transformations are not always perceived as positive or desirable, particularly when they pose a threat to dominant values and beliefs or are imposed by others with interests, influence, and power. History is full of examples where imposed transformations have failed, resulting in resistance or ruin in empires, states, companies, or communities. The signs and signals of transformation can sometimes be difficult to detect, except in retrospect. Futurists – people who pay attention to larger societal trends – often look to emerging innovations in science and technology (Riedy 2009). Recent and ongoing changes that may be signs of transformations include 3-D printing technologies, the proliferation of smartphones, development of artificial intelligence, and the rise of virtual-reality experiences. However, the types of transformation needed to address climate change and sustainability challenges will involve more than new gadgets and experiences. They are likely to also involve "interior" changes in worldviews, values, or paradigms that manifest as new ways of relating to others, treating nature, and organizing society. Recent examples include more flexible attitudes towards gender and sexuality, greater attention to social justice, and a growing recognition of animal rights.

Yet even transformations that are aimed at enhancing sustainability, tolerance, inclusion, social connection, and well-being can be unsettling. For example, the invention of the 3-D printer has the potential to reduce demand for transport of physical goods and contribute to more sustainable business practices, but it may also disrupt global supply chains and severely reduce the need for unskilled labor (Mohr and Khan 2015). Transformations may create instabilities that can lead to surprises, uncertainties, disappointments, or anxieties, even among those who are eager for change. Because transformations are seldom predictable and manageable, many people experience them as stressful or disruptive. They can be particularly threatening to those who have a stake or strong interest in maintaining the status quo. As Mark Pelling (2011) points out, this includes both those in power and also those who might benefit from change but are wary of the additional instabilities and uncertainties that accompany transformations. Not surprisingly, transformations often create resistance, counter-pressures, and backlashes that can block change.

Limiting global warming to 2°C – and ideally to 1.5°C – by the end of this century will without a doubt involve significant transformations (O'Brien 2018). However, what needs to be transformed, how it should be transformed, and who is responsible for initiating transformation is far from clear. As with other facets of climate change, the types of transformations that are considered necessary largely depend on how the problem is framed. Among the discourses that accept the scientific evidence of climate change, biophysical approaches often emphasize transformation through innovations in renewable energy, regulations of fossil fuel energy production, and international collaboration. Critical approaches suggest that political, social, and economic transformations are needed to address the root causes of environmental degradation and social vulnerability. Integrative approaches recognize the importance of both practical and political changes, yet also emphasize a need for deeper changes, such as rethinking and reimagining how humans are connected to each other and to nature. Although each framing prioritizes different types of strategies to address climate change, a common and cross-cutting theme is a shared recognition of both the scope of the climate change problem and the urgent need for deliberate transformations to a more sustainable world. In the next section, we discuss an integrative approach to transformation that reveals numerous entry points for addressing climate change.

Transformations to sustainability: An integrative approach

The concept of transformation can be understood from many perspectives, as evidenced in a large and growing body of research from the natural and social sciences and humanities (Feola 2015; Fazey et al. 2018). This includes research on social-ecological transformations, socioeconomic transformations, socio-technical transitions, as well as on the role of narrative, creativity, imagination, and the arts in transformation processes (EEA 2017; Galafassi et al. 2018). While there are diverse approaches and many entry points to transformation, the concept itself can be more generally defined as significant changes in form, structure, and/or meaning making (O'Brien 2018). Transformation represents more than linear incremental shifts from one state or condition to another; it is often characterized by non-linear changes where small perturbations lead to rapid shifts (see Figure 10.1). This can include subtle, qualitative shifts that change the characteristics of a system or catalyze the emergence of new properties within complex systems (Chesters and Welsh 2005). As discussed above, transformations often challenge existing patterns and "business as usual," and upset the status quo.

The multidimensional nature of transformation processes can be represented through the heuristic of three spheres. This simple model considers transformation as a continuous process that involves interrelated changes across three dimensions, referred to as the practical, political, and personal spheres (IPCC 2018b; O'Brien 2018) (see Figure 10.2). These three interacting spheres depict how changes in form, structure, and meaning making together contribute to transformation processes. The model recognizes that transformation is a multifaceted process that cannot be limited or reduced to only one dimension. It also implies that there are multiple entry points for engaging with deliberate transformations towards an equitable, sustainable, and thriving world.

The most common entry point for engagement is the practical sphere of transformation, which focuses on sustainability outcomes through changes in form. Actions and activities in the practical sphere are often aimed at the realization of measurable results and goals, such as lower carbon

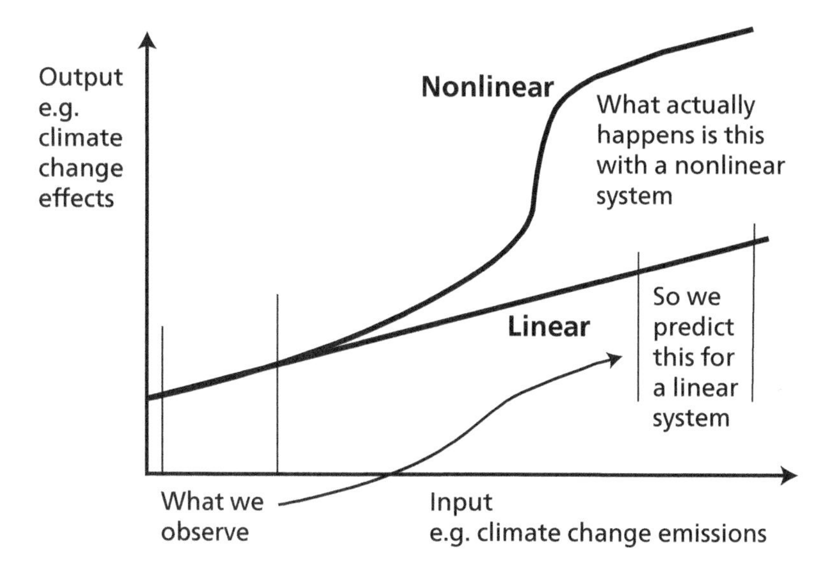

Figure 10.1 Linear and non-linear change.
Source: Karen Gardner

dioxide emissions or climate change adaptation. They may include efforts to install more solar panels and windmills, reduce meat consumption, manage development in flood-prone regions, and so on. Transformations in the practical sphere frequently involve innovations in technologies and changes in behaviors and practices that lead to observable outcomes. Their feasibility is often related to appropriate investments in research and development, improved management, transfer of technology, or policy incentives to support behavioral change.

Although practical transformations are fundamental for addressing climate change, they need to be supported by structural and systemic changes. This draws attention to the political sphere of transformation which includes social and cultural norms, institutions, and governance systems that shape behaviors, actions, and investments. It is in the political sphere where the "rules of the game" are negotiated and decided, where goals and outcomes are prioritized, and where conflicts and movements emerge to directly or indirectly influence systems. For example, regulations or taxes that encourage energy savings or the development of renewable energy resources must be negotiated and implemented amidst competing political commitments, priorities, and goals, such as maintaining jobs in fossil fuel-producing regions. Transformations in the political sphere refer to more than changes in formal politics, although these may be important in some contexts.

The personal sphere of transformation represents changes in meaning making, which includes the individual and shared beliefs, values, worldviews,

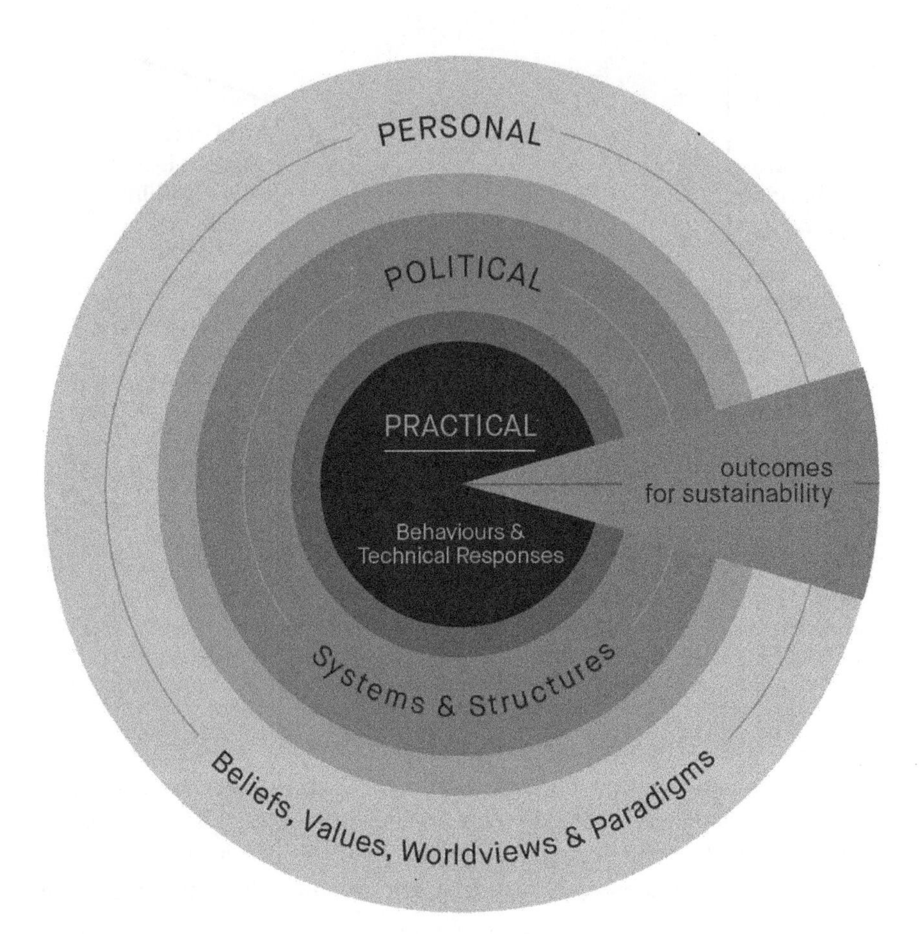

Figure 10.2 Three spheres of transformation.
Source: Based on O'Brien and Sygna 2013

and paradigms that shape attitudes, actions, and perceived options. These "subjective" dimensions influence preferred strategies or approaches to transformation, including perceptions of individual and collective agency. Beliefs, values, and worldviews can explain why certain actions in the practical and political spheres are prioritized while others are dismissed or ignored. The personal sphere implicitly influences the goals or objectives of systems, including who can or should benefit from them, and whose decisions count. It draws attention to the social and cultural construction of meaning, which influences rules, norms, and behaviors, as well as what systems and structures are considered to be fixed or unchangeable. It also acknowledges the importance of social consciousness in transformation processes (Schlitz, Vieten, and Miller 2010).

While the three spheres may appear to be separate, it is important to keep in mind that this is merely an abstraction; in everyday life, we experience continuous connections and interactions between the three spheres. Transformations in the practical sphere (for example, a pronounced reduction in food waste) can be supported by transformations in the political sphere (for example, creating institutions that facilitate investments in food storage facilities or changes in cultural norms related to "use before" dates). Changes in systems and structures can in turn be influenced by transformations in the personal sphere (for example, changing beliefs about leftover foods). By linking the practical, political, and personal spheres of transformations, we see how outcomes for sustainability entail social processes that typically involve all three dimensions (see Box 10.1). In the next section, we describe some ongoing efforts and actions to promote transformations to sustainability and emphasize the potential to accelerate these shifts by recognizing and activating interconnected changes among all three spheres of transformation.

Transformations are happening

To change at the rate, scale, scope, and depth needed to meet the targets of the Paris Agreement while also addressing the root causes of risk and vulnerability calls for deliberate transformations. Some people assume that it is too late and that nothing can be done. However, we do have a large body of both knowledge and experience with transformational change, much of which comes from the social sciences (Hackmann, Moser, and St Clair 2014; O'Brien 2018). In fact, efforts to promote transformations to sustainability have been underway for many decades and are increasingly visible (Bennett et al. 2016). Recognizing the many facets of transformation, a key challenge is to work skillfully to create the conditions for individuals and groups to both transform and thrive.

An early call for transformation to sustainability can be traced back to the 1987 Brundtland Commission Report on "Our Common Future," which popularized the concept of sustainable development. This report recognized the importance of both people and environment, including the rights of future generations (WCED 1987). In recent years, transformations to sustainability have been promoted through **green transitions**, with a strong emphasis on energy production and usage, consumption, and transportation. Meanwhile, climate change activists have also been drawing attention to tensions between development-as-usual, equity, and sustainability, underscoring the need for **just transitions** (Swilling and Annecke 2012). Still others are making appeals for entirely new paradigms and ways of thinking about the relationships between humans, non-humans, and the environment (Bennett 2010; Wendt 2015). Below, we explore some of these efforts and consider their shared potential for catalyzing transformations to sustainability.

Box 10.1 Exploring your own role in transformations to sustainability

Who actually decides which futures we will collectively experience? Whose responsibility is it to take action? These are difficult questions, and many assume the answers to be "someone else." Some people will point to oil companies, banks, and multinational corporations as having the greatest responsibility for taking action to address climate change, while others say that it is the role of politicians, CEOs, or spiritual leaders. In democratic societies, many argue that "we the people" collectively decide the future. More and more attention is given to the attitudes and visions of young people, who arguably have the biggest stake in the future (Ojala 2012; Carabelli and Lyon 2016).

What role will you play in transforming the future? In the practical sphere, what can you do to contribute to sustainable outcomes? How can you lower your carbon footprint? This may mean reflecting on your own habits and practices, and seeing how they relate to social norms and the ways that society is organized. For example, are there options for public transportation and infrastructure to support walking or bicycling? Is it acceptable in your social circles to eat less meat, and, if not, why? How do your actions influence and inspire others?

In the political sphere, what systems and cultures do you think need to be shifted? What alternatives to dominant systems would you like to support or develop? How can you influence these shifts? Where can your voice make a difference? What types of groups can you create or join to have a greater influence? Actions might involve advocating for renewable energy policies, engaging in the sharing economy, or coming up with something totally new. Can you see yourself becoming active in politics?

In the personal sphere, what are your beliefs about change? Do you think you can make a difference? If not, what are some of the beliefs and assumptions that limit your engagement? What values do you stand for and why? Can you discuss these values with your family, friends, and community? What are your ideas about leadership? As Monica Sharma discusses in *Radical Transformational Leadership*, every person has the potential to be a game-changer and make a difference: "As human beings, we have an innate ability to commit to action when we get out of our own way. Committing to action is what makes ordinary people do extraordinary things" (2017: 59). How can your passions, skills, talents, and strengths be best leveraged to contribute to transformations to sustainability?

Green transitions

In many countries, cities, and communities, there is a growing emphasis on the idea of green transitions. This is an umbrella term that covers concepts such as green energy transitions, green mobility transitions, green consumption transitions, and green urban transitions. Strategies for green transitions typically favor techno-managerial and behavioral solutions that can be developed, implemented, monitored, and evaluated, often with the objective of meeting specific targets and goals, such as the Paris Agreement or Sustainable Development Goals (SDGs). Green energy transitions might entail a shift from coal-based heating to renewable sources, the retrofitting of old buildings to make them more energy efficient, or the development of passive housing units that use minimal energy for heating or cooling. Green mobility transitions may involve lowering CO_2 emissions through more fuel-efficient automobiles, buses, and airplanes or by encouraging bicycling, walking, and the use of public transport. Green consumption transitions may focus on eliminating plastic packaging, fewer purchases of fast fashion, or reusing and recycling household waste. All of these approaches involve practical actions that produce measurable results, particularly if backed by supportive policies.

The language of green transitions speaks to the business community, social innovators, policy makers, and others who need to deliver measurable results yet are often constrained by commitments to maintaining or creating jobs, profits, or economic growth. Decisions and actions taken in the business sector play an important role in efforts to promote green transitions, particularly within sectors associated with high energy use, such as transport, buildings, and consumer products. Such decisions can entail changes in design, the types of materials and energy that are used in production processes, sources of input materials, and how products are packaged and delivered. A number of companies have signed on to the United Nations Global Compact, a voluntary initiative that commits companies to implementing sustainability principles consistent with the SDGs and the Paris Agreement (UNGC 2018). Many companies have also begun to self-disclose carbon emissions through initiatives such as the Carbon Disclosure Project (CDP 2018). Indeed, "going green" can be an effective marketing tool for some businesses or a response to civil society activism for others, leading to selective disclosure that can be either symbolic or substantive (Marquis, Toffel, and Zhou 2016). Dangelico and Vocalelli (2017) review how the green marketing concept has changed over time to become a strategy that takes into account broader global sustainability issues, not just by promoting environmentally friendly products but by branding the image of the company as "green."

The food and agricultural sectors are relevant for many efforts to support green transitions. As we discussed in chapter 5, increasing consumption of plant-based foods and reduction of deforestation can play important roles in mitigating climate change. For example, the idea of climate-smart agriculture (CSA) has been promoted by the UN Food and Agriculture Organization as an innovative means of supporting livelihoods and food security, emissions reductions, and climate change adaptation (FAO 2016). International agricultural research centers, as well as agri-business firms, are promoting CSA through the development of new seeds, digital tools, fertilizers, and cropping methods to help both increase carbon sequestration and adapt to changing climate conditions. Sustainable intensification, another cornerstone of CSA, has been promoted to enhance smallholder livelihoods through strategies such as banana–coffee intercropping and agroforestry-based livestock diets (Campbell et al. 2014). Although CSA and related approaches to green agricultural and land-use transitions are described as "win–win" solutions that will lead to greater agricultural productivity, improved food security, increased biodiversity, and better health outcomes, in practice they sometimes involve adoption of costly technologies that are unaffordable for small farmers, contribute to the expansion of industrial agriculture, or result in land grabbing (Zoomers 2010; Clapp, Newell, and Brent 2018). Avoiding such outcomes may require changes within the political sphere such as transformation of governance systems in order to facilitate greater local control over resources (Patterson et al. 2017).

Efforts to support green transitions often hinge on the added value or co-benefits of going green. Recall that co-benefits are positive additional consequences that result from particular policies and actions. For example, bicycling is not only a climate-friendly mode of transport but has co-benefits that might include saving money on car maintenance or public transport, getting in better shape, losing weight, feeling more relaxed and less stressed, or simply enjoying the outdoors. In the case of diets, Springmann et al. (2016: 4146) found that "Transitioning toward more plant-based diets that are in line with standard dietary guidelines could reduce global mortality by 6–10% and food-related greenhouse gas emissions by 29–70% compared with a reference scenario in 2050." Efforts to reduce greenhouse gas emissions can have co-benefits of reduced air pollution, energy-cost savings from lower fuel usage, and improved public health (Lelieveld et al. 2015). China's recent efforts to reduce urban air pollution from coal-burning power plants and industrial machinery through stricter regulation of emissions have contributed to improved air quality while also reducing greenhouse gas emissions (Jiang et al. 2013).

Green transitions call for some degree of environmental awareness and engagement, thus they often include a strong information element that tends to be oriented around projects, campaigns, labeling, and other initiatives. Recognizing that awareness and engagement are often lacking, green transitions have also been promoted through nudging people to make the right decisions by making such choices easy and obvious (e.g., providing smaller plates at restaurants to reduce food waste). Nudging is a subtle approach to behavioral change and can be considered cost-effective; thus it tends to be a favored approach by the public sector. However, nudging can also be considered manipulative and unethical, and it may do little to alter underlying attitudes towards the environment or create long-term change (Schubert 2017). Lieberoth, Holm Jensen, and Bredahl (2018), for example, studied the role that nudging and other **soft measures** play in getting automobile users to use public transportation in Denmark and reported that they have limited effects. They found that these efforts need to be accompanied by larger-scale policies and infrastructural changes such as road taxes, bike lanes, and increased and subsidized public transport.

Many transition initiatives focus on reworking or improving present systems, making them more energy efficient and effective – in other words, *cleaner*, *greener*, and *smarter*. Although an emphasis on green transitions captures the urgent need to shift to sustainable practices, it risks both overstating the role of technologies and reducing the problem to individual or corporate behaviors, rather than recognizing sustainability as a collective challenge tied to the structures of larger social, economic, and political systems. For example, reliance on the business sector for emissions reductions in response to the urgency of climate change action can downplay the role of market imperatives and instead support business-as-usual approaches to sustainability (Wright and Nyberg 2016). Some of these approaches have been criticized for supporting prevailing social, economic, or gender relations of power and inequality and providing no coherent argument for large-scale change (Warner 2010; Standal and Winther 2016) (see Box 10.2). Ensuring the effectiveness of sustainability efforts often calls for substantive changes in the political and personal spheres, including some of the foundational assumptions about economic, social, and human–environment relationships.

Practical solutions associated with green transitions seldom account for the ongoing impacts of climate change. Although many sustainability efforts focus on both mitigation and adaptation, practical strategies for reducing greenhouse gas emissions often include an implicit assumption that climate conditions, resource availability, and settlement patterns will be similar in the future. Few transition initiatives explicitly recognize that the consequences of climate change impacts in other places will have ripple effects, whether through displaced populations, disruptions of global production chains, or degradation of ecological systems. Accounting for climate change impacts and their effects across scales will be critical for successful green transitions.

Box 10.2 Green growth and green economies

The theory of ecological modernization underpins many of the efforts aimed at green transitions. Ecological modernization is a school of thought that considers environmental transitions as beneficial to the economy (Mol and Spaargaren 2000). The idea that the economic and the environmental aspects of development can be successfully combined has given rise to calls for green growth and promotion of the green economy as a new type of development. The concept of leapfrogging is sometimes raised as an alternative development pathway – this refers to the idea that low-income economics can bypass inefficient and polluting technologies by adopting clean and green solutions right away (van Benthem 2015).

The United Nations Environment Programme is championing the green economy initiative as an alternative to the current economy, highlighting its potential to promote human well-being and social equity while at the same time reducing environmental risks and ecological scarcities (United Nations Environment Programme 2018). A low-carbon, resource-efficient, and socially inclusive economy is envisioned to be one where growth in income and employment is driven by investments that address climate change, biodiversity loss, and reductions in ecosystem services. Such investments will rely on targeted public expenditures, policy reforms, and regulatory changes. There are, however, a number of challenges to pursuing and funding a green economy, particularly in light of biodiversity and ecosystem degradation (Barbier 2011). Moreover, the goal of continuous economic growth, green or not, is considered by some to be the ultimate source of many environmental problems as it demands new markets, new products, and increasing resource use and consumption.

Do you think that the idea of a green economy is practical and feasible? Does this approach appeal to you? Why or why not? What concerns does this approach raise?

Climate change activism and social movements

Climate change activism and social movements are rooted in the political sphere of transformation, which recognizes the role of power, politics, interests, and agency. Climate change activism often stems from disagreement and a desire to challenge a prevailing view, policy, practice, decision, institution, idea, or assumption that is seen as contributing to climate change (O'Brien, Selboe, and Hayward 2018). These approaches may entail local-scale, community-based actions, including awareness-raising events, educational programs, and sustainability campaigns. They may also include involvement in global organizations aimed at promoting climate action, such as 350.org, Global Power Shift, Idle No More, La Via Campesina, and the Youth Climate Movement. Political engagement can also occur in more individualized or specialized ways, such as issue-specific activism or low-threshold and part-time activities, e.g., promoting recycling or bicycling, signing petitions on social media, or consuming environmentally friendly products (O'Brien, Selboe, and Hayward 2018). Many climate actions are intended to challenge business-as-usual economic and social policies, including their emphasis on economic growth (Escobar 2015). These include divestment campaigns, boycotts, legal actions, and other means of shifting political and economic power away from the fossil fuel industries and carbon polluters (O'Brien, Selboe, and Hayward 2018).

A number of climate-related social movements fall under the umbrella of climate justice. The concept of climate justice, which is closely related to the idea of environmental justice, draws attention to a wide range of ethical, justice, and equity concerns surrounding climate change (Bulkeley et al. 2013). These include the equity dimensions of greenhouse gas emissions and those associated with uneven climate change impacts, vulnerability, and adaptive capacity, as discussed in earlier chapters. Movements for climate justice also draw attention to the linkages between green transitions, rights for workers, and job creation (Hampton 2018). Job creation in sectors such as solar and wind energy, green buildings, recycling and reuse, and so forth, is a widely discussed co-benefit of many types of green transitions. According to the International Labour Organization (ILO), green jobs are "decent jobs" that "reduce the consumption of energy and raw materials, limit greenhouse gas emissions, minimize waste and pollution, protect and restore ecosystems and enable enterprises and communities to adapt to climate change" (ILO 2018: 53). Yet many jobs in the industries that promote green transitions are not necessarily green from the perspective of the person doing the job. Production of solar panels or recycling of electronics can expose workers to dangerous toxins. Sorting garbage or working on an organic farm can be dangerous and physically demanding yet provide minimal pay.

In response to concerns over working conditions and pay in green industries and job losses in traditional industries (e.g., fossil fuel mining and production), the ILO and other organizations have embraced just transitions. This concept captures the idea that transitions away from a fossil fuel economy need to incorporate principles of energy and environmental justice (Heffron and McCauley 2018). Ensuring that green transitions are just and fair might mean taking steps to support workers and improve working conditions in affected industries. Alternative employment opportunities, income support, retraining, assistance in relocating, and so forth, might all be part of a just transition strategy. At a broader scale, the just transition movement seeks to address equity issues that may arise from an effort to ensure that "going green" does not contribute to increased poverty and inequality or lead to spatial displacement of highly polluting industries to poorer countries. It also recognizes that justice is critical to climate change adaptation (Schlosberg, Collins, and Niemeyer 2017).

There are diverse ways that climate activism can be expressed, representing different approach to change, different relationship to power, and different ways of expressing dissent (see Box 10.3) (O'Brien, Selboe, and Hayward 2018). Often social and environmental activism promoting equitable and sustainable transformations remains unseen or "off the radar" in the global media. Yet, as Paul Hawken (2008) writes in *Blessed Unrest*, we are in the middle of a large and unstoppable movement to reimagine our relationship to the environment and to one another. This bottom-up, unorganized movement has largely been invisible, a common characteristic of non-linear processes of social transformation.

Shifting paradigms

The idea of shifting paradigms, or shared ways of thinking, represents another opening for transformation. Returning to chapter 4's discussion of worldviews and views of nature, recall that how we see nature (whether as separate from or integrated with humanity) contributes to how we understand the problem of climate change and the types of solutions that are proposed. Paradigm shifts are considered to be a particularly powerful leverage point for systems change (Abson et al. 2017; Göpel 2016; O'Brien 2018). Innovative technologies, ideas, and modes of social organization have emerged throughout history as the result of individuals and groups challenging assumptions, defying conventions and norms, and realizing alternative visions. As Meadows (1999: 18) writes, "Paradigms are the sources of systems . . . from shared social agreements about the nature of reality[,] come system goals and information flows, feedbacks, stocks, flows and everything else about systems."

In identifying opportunities for paradigm shifts, it is useful to remember that what is "normal" has changed significantly over time, whether in relation to gender roles, social relationships, eating habits, hygiene, teaching methods, attitudes towards children, and, not least, attitudes towards nature. Nonetheless, there is still a diversity of norms associated with different cultures, age cohorts, and social groups. As discussed in earlier chapters, norms and practices associated with the habits of capitalism, including high-energy lifestyles, hypermobility, and a belief in continuous economic growth, are key drivers of climate change (Wilhite 2016). Transformative responses to climate change may require challenging the culture of automobiles, diets that are high in meat and dairy products, fast fashion, binge travel, single-use

Box 10.3 Climate activism among youth

Young people have a critical stake in policies and actions that will shape the future climate. In a study of youth climate activism in high-emission countries, O'Brien, Selboe, and Hayward (2018) identify three major forms of youth activism and engagement with climate change, which they identify as "dutiful dissent," "disruptive dissent," and "dangerous dissent."

Dutiful dissent entails joining activities within mainstream political parties and environmental organizations that support a variety of approaches and actions for reducing greenhouse gas emissions or facilitating adaptation. Dutiful actions work through existing political and economic institutions in ways that strengthen and uphold their legitimacy. This type of dissent represents resistance to the status quo, yet it also adheres to the "script" of current institutions, hegemonic powers, and economic systems. Rather than challenge existing political decision processes, it remains dutiful to their logic and existence. As such, it tends to promote a depoliticized and post-political response to climate change that frames it in consensual and technocratic terms (Swyngedouw 2013).

Disruptive dissent seeks to modify or change existing political and economic structures, including the norms, rules, regulations, and institutions that maintain them. Disruptive actions are targeted towards challenging power relationships, as well as the actors and political authorities who maintain them, often through direct protests and collective organization. Such actions may involve starting or joining boycotts, disrupting international climate meetings to draw attention to hypocrisy or exclusion of important voices, or participating in protest marches and political rallies. These vocal challenges to the status quo can open spaces for new actors, but they also present a polarizing and antagonistic vision which can limit mobilization and inhibit new ideas and alternatives views of the future (O'Brien and Selboe 2015; Kenis and Lievens 2014).

Dangerous dissent is a type of political activism that defies business as usual by initiating, developing, and actualizing alternatives that have the potential to inspire and sustain long-term transformations. They can include a wide spectrum of actions, ideas, discourses, practices, tactics, alliances, and technologies. Such approaches may not even be labeled as climate change activism. They might include, for example, adopting a plant-rich diet or living "off the grid." Such actions have the potential to influence social norms and undermine interests that are complicit in maintaining current systems of unsustainable growth, high greenhouse gas emissions, and deep social injustice (O'Brien, Selboe, and Hayward 2018). Dangerous dissent also challenges existing paradigms and worldviews, which can lead to the transformation of systems.

Have you ever engaged in any form of activism? If so, do you consider your engagement to be dutiful, disruptive, or dangerous to business as usual and the status quo? Do you think that activism is likely to be an effective strategy for addressing climate change?

plastics and throwaway goods, and other tenets of the consumption–growth paradigm. Sarah Strauss reflects on the challenge at hand:

> If we take seriously our culture's full dependence on petroleum-based products, we come quickly to the realization that addressing greenhouse gas production is tied to a massive shift in not only the forms of energy we use for transportation, manufacture, and heating, but also therefore a massive shift in the ways of relating to the rest of the world – socially, culturally, environmentally – that is at least as dramatic as that which the Greenlandic hunters of the north are facing. (Strauss 2012: 4)

Paradigm shifts often occur in times of crisis or as new generations question traditions, precedents, and givens. In Thomas Kuhn's 1962 book *The Structure of Scientific Revolutions*, he pointed out that "probably the single most prevalent claim advanced by proponents of a new paradigm is that they can solve problems that have led the old one to a crisis" (Kuhn 1962: 135). Climate change, biodiversity loss, rising inequality, and other persistent global problems suggest that current paradigms are in crisis and that new paradigms are needed. New paradigms, however, are already available, and they are generating many of the transformations that are currently underway. For example, Andreas Weber's (2013) "enlivenment" paradigm emphasizes the importance of humanity's actual, lived, and felt experience and its connection with a nature that is alive and creative. Alexander Wendt's (2015) quantum social theory recognizes the role of consciousness in social processes. Jane Bennett's "vibrant matter" considers the role of vital agency in non-human things. These and many other emerging paradigms are challenging dominant views of human–environment relationships, showing that we are connected and that our ideas and actions matter more than we think. More important, they are contributing to a growing sense of transformative agency (Westley et al. 2013).

As the integrative discourse reminds us, transformations to sustainability are not only cognitive but embodied. A "head–heart–hand" approach to transformations recognizes the importance of critical reflection about climate change and its drivers and consequences. It also acknowledges the power of a passionate commitment to equity and justice in both processes and outcomes, Finally, it emphasizes the significance of human agency and collective action, whether through social movements, innovations, or leadership, to foster new ways of doing things, organizing societies, and thinking about the future.

Transforming the future

There are many ways to think about the future. Our thoughts, ideas, and expectations about the future are never neutral – they influence what we do or do not do, how we spend our time, energy, and resources, and what we leave for future generations. Narratives, or storylines, are powerful because we often enact or live our stories, creating the reality we expect to experience (Wright 2010; El Khoury 2015). This means that how we perceive climate change and its solutions, including our own role in realizing them, is of profound significance.

With climate change, there is often a tendency to tell stories about the future as dystopic. The dystopian future under climate change is generally characterized by worsening environmental conditions and social disruptions that lead to more and more catastrophes, and eventually to the collapse of social-ecological systems (Kaplan 2015). Dystopic futures are sometimes easier to imagine than utopian alternatives because they do not require us to change. We can simply extrapolate from the past into the future by picturing a world with warmer temperatures, changed rainfall patterns, sea-level rise, and ocean acidification, then watch our expectations play out over time. A dystopic future may also be a future of two worlds: an eco-modernist world that is clean, green, smart, and resilient, existing alongside a disrupted world characterized by displaced populations living in poverty, degradation, and misery. Such a trajectory would be unacceptable for many people and in fact may ultimately be impossible to maintain, given that climate change is a systemic risk that will be experienced globally.

Utopias, in contrast, refer to places or states of existence representative of the "good society" (Hjerpe and Linnér 2009). Imagine, for example, nine billion people thriving on a planet that provides food, water, energy, and resources to satisfy people's needs, where difference is welcomed or tolerated, rather than judged or condemned, and where diversity is celebrated and all people and species have recognized rights and acknowledged responsibilities to each other and to the future. Although such visions may be appealing, Erik Olin Wright (2010) argues that we need "real utopias" that embrace the tension between dreams and practice. Real utopias will require transformations, and many of these may be difficult or disruptive, particularly if they challenge vested interests or business as usual. In the context of climate change, utopian visions for society have to include not only mitigation and adaptation but also a wide range of transformations that protect the things that we value, including nature as we know it today, in an equitable and ethical way.

As we look to the future, perhaps there is a silver lining in anthropogenic climate change. In contrast to volcanic eruptions, meteor strikes, variations in the Earth's orbit, sunspot activity, or other processes and events that influence the climate system, human activities can be consciously transformed. The future that we experience will, to a large extent, depend on the actions and decisions that are taken today, tomorrow, and in the coming years. Depending on the context, it is said that transformations are catalyzed when only about 10–25% of a group share a strong belief – not 51% or 100% (Xie et al. 2011; Centola et al. 2018). This is because people are connected through language, stories, and shared meaning. Every person inspires others and can be inspired by the ideas, words, and actions of others. As more and more people realize that they do matter and can make a difference in the practical, personal, and political spheres, transformations that are already underway will accelerate and become more visible.

Summary

Throughout this book, we have highlighted climate change as a social issue, and we have emphasized that people are, in fact, the most powerful solution to climate change. Not only do we develop, use, and manage the technologies that can benefit or harm the environment, but we also have the capacity to critically reflect, engage, and work together for positive change. The role of the individual mirrors the role of the collective, and positive changes have the potential to accelerate remarkably as more people recognize that climate change is not simply an environmental problem but a social problem. Because climate change is driven by human activities, it provides many openings for us to make individual and collective choices about the type of future that we desire and the pathways that will take us there. By recognizing the social dimensions of climate change, we open new opportunities for transformation to a more equitable, sustainable, and thriving future.

Reflection questions

1 Consider a historical example or recent event where social conditions changed very rapidly. How did this change come about? Does this example offer any lessons for addressing climate change?
2 What are some examples of dangerous dissent, or climate activism that defies business as usual? Can you think of actions that might inspire and sustain long-term transformations?
3 Futurist and inventor R. Buckminster Fuller posed the question “If the success or failure of this planet, and of human beings, depended on how I am and what I do, how would I be? What would I do?” Have you considered your role in addressing climate change? How you can extend your role beyond your current sphere of influence?

Further reading

Abson, D. J., Fischer, J., Leventon, J., et al., 2017. Leverage points for sustainability transformation. *Ambio* 46(1), 30–9.

Göpel, M., 2016. *The Great Mindshift: How a New Economic Paradigm and Sustainability Transformations go Hand in Hand*. Cham, Switzerland: Springer International.
O'Brien, K., 2018. Is the 1.5°C target possible? Exploring the three spheres of transformation. *Current Opinion in Environmental Sustainability* 31: 153–60.
Sharma, M., 2017. *Radical Transformational Leadership: Strategic Action for Change Agents*. Berkeley, CA: North Atlantic Books.
Swyngedouw, E., 2013. The non-political politics of climate change. *An International Journal for Critical Geographies* 12: 1–8.

References

Aasness, M. A. and Odeck, J., 2015. The increase of electric vehicle usage in Norway: Incentives and adverse effects. *European Transport Research Review* 7(4): 34.

Abson, D. J., Fischer, J., Leventon, J., et al., 2017. Leverage points for sustainability transformation. *Ambio* 46(1): 30–9.

Adger, W. N., 2006. Vulnerability. *Global Environmental Change* 16(3): 268–81.

Adger, W. N., Barnett, J., Brown, K., Marshall, N., and O'Brien, K., 2012. Cultural dimensions of climate change impacts and adaptation. *Nature Climate Change* 3(2): 112–17.

Adger, W. N., Dessai, S., Goulden, M., et al., 2009. Are there social limits to adaptation to climate change? *Climatic Change*, 93 (3–4): 335–54.

Adger, W. N., Pulhin, J. M., Barnett, J., et al., 2014. Human security, in C. B. Field, V.R. Barros, D. J. Dokken, et al. (eds), *Climate Change 2014: Impacts, Adaptation, and Vulnerability. Part A: Global and Sectoral Aspects. Contribution of Working Group II to the Fifth Assessment Report of the Intergovernmental Panel on Climate Change.* Cambridge: Cambridge University Press, 755–91.

Afifi, T., 2011. Economic or environmental migration? The push factors in Niger. *International Migration* 49(s1): e95–e124.

Afionis, S., Sakai, M., Scott, K., Barrett, J., and Gouldson, A., 2017. Consumption-based carbon accounting: Does it have a future? *Wiley Interdisciplinary Reviews: Climate Change* 8(1): e438.

Agbongiarhuoyi, A. E., Abdulkarim, I. F., Fawole, O. P., Obatolu, B. O., Famuyiwa, B. S., and Oloyede, A. A., 2013. Analysis of farmers' adaptation strategies to climate change in cocoa production in Kwara State. *Journal of Agricultural Extension* 17(1): 10–22.

Ahmed, S. and Stepp, J. R., 2016. Beyond yields: Climate change effects on specialty crop quality and agroecological management. *Elementa: Science of the Anthropocene* 4 (000092).

Ajibade, I., McBean, G., and Bezner-Kerr, R., 2013. Urban flooding in Lagos, Nigeria: Patterns of vulnerability and resilience among women. *Global Environmental Change* 23(6): 1714–25.

Alaimo, S., 2010. *Bodily Natures: Science, Environment, and the Material Self.* Bloomington, IN: Indiana University Press.

Albrecht, G., Sartore, G.-M., Connor, L., et al., 2007. Solastalgia: The distress caused by environmental change. *Australasian Psychiatry: Bulletin of Royal Australian and New Zealand College of Psychiatrists* 15 (Suppl. 1): S95–8.

Aldy, J. E. and Stavins, R. N., 2012. The promise and problems of pricing carbon: Theory and experience. *The Journal of Environment & Development* 21(2): 152–80.

Alkire, S., 2003. *A Conceptual Framework for Human Security.* Centre for Research on Inequality, Human Security and Ethnicity, CRISE. University of Oxford, Working Paper 2.

Allison, E. A., 2015. The spiritual significance of glaciers in an age of climate change. *Wiley Interdisciplinary Reviews: Climate Change* 6(5): 493–508.

Alston, M., 2011. Gender and climate change in Australia. *Journal of Sociology* 47(1): 53–70.

Altvater, E., Crist, E. C., Haraway, D. J., Hartley, D., Parenti, C., and McBrien, J., 2016. *Anthropocene or Capitalocene? Nature, History, and the Crisis of Capitalism.* Oakland, CA: PM Press.

Ambrose, K., 2017. Cherry blossom tracking: Saturday's peak bloom caps a strange season on the Tidal Basin. *Washington Post*, March 27.

Anderson, E. P., Jenkins, C. N., Heilpern, S., et al., 2018. Fragmentation of Andes-to-Amazon connectivity by hydropower dams. *Science Advances* 4(1): eaao1642.

Anderson, K., 2015. Duality in climate science. *Nature Geoscience* 8(12): 898–900.

Anderson, Z. R., Kusters, K., McCarthy, J., and Obidzinski, K., 2016. Green growth rhetoric versus reality: Insights from Indonesia. *Global Environmental Change* 38: 30–40.

Ang, B. W., Choong, W. L., and Ng, T. S., 2015. Energy security: Definitions, dimensions and indexes. *Renewable and Sustainable Energy Reviews* 42: 1077–93.

Anguelovski, I., Shi, L., Chu, E., et al., 2016. Equity impacts of urban land use planning for climate adaptation: Critical perspectives from the global north and south. *Journal of Planning Education and Research* 36(3): 333–48.

Ansar, A., Caldecott, B. L., and Tilbury, J., 2013. *Stranded assets and the fossil fuel divestment campaign: What does divestment mean for the valuation of fossil fuel assets?* University of Oxford: Smith School of Enterprise and the Environment.

Anthony, K. R. N., 2016. Coral reefs under climate change and ocean acidification: Challenges and opportunities for management and policy. *Annual Review of Environment and Resources* 41(1): 59–81.

Ashenfelter, O. and Storchmann, K., 2016. The economics of wine, weather, and climate change. *Review of Environmental Economics and Policy* 10(1): 25–46.

Bain, P. G., Hornsey, M. J., Bongiorno, R., and Jeffries, C., 2012. Promoting pro-environmental action in climate change deniers. *Nature Climate Change* 2(8): 600–3.

Barad, K., 2007. *Meeting the Universe Halfway: Quantum Physics and the Entanglement of Matter and Meaning.* Durham, NC: Duke University Press.

Barbier, E., 2011. The policy challenges for green economy and sustainable economic development. *Natural Resources Forum* 35(3): 233–45.

Barnes, M., 2015. Transit systems and ridership under extreme weather and climate change stress: An urban transportation agenda for hazards geography. *Geography Compass* 9(11): 604–16.

Barnett, J. and Adger, W. N., 2007. Climate change, human security and violent conflict. *Political Geography* 26(6): 639–55.

Barnett, J. and Campbell, J., 2010. *Climate Change and Small Island States: Power, Knowledge, and the South Pacific.* Washington, DC: Earthscan.

Barnett, J. and O'Neill, S., 2010. Maladaptation. *Global Environmental Change* 20(2): 211–13.

Barnett, J., Evans, L., Gross, C., et al., 2015. From barriers to limits to climate change adaptation: Path dependency and the speed of change. *Ecology and Society* 20(3): 5.

Barnett, J., Lambert, S., and Fry, I., 2008. The hazards of indicators: Insights from the environmental vulnerability index. *Annals of the Association of American Geographers* 98(1): 102–19.

Bassett, T. J. and Fogelman, C., 2013. Déjà vu or something new? The adaptation concept in the climate change literature. *Geoforum* 48: 42–53.

BBC, 2015. Deadly floods hit Chennai and Sri Lanka. *BBC News*, November 17.

Bennett, E. M., Solan, M., Biggs, R., et al., 2016. Bright spots: Seeds of a good Anthropocene. *Frontiers in Ecology and the Environment* 14(8): 441–8.

Bennett, J., 2010. *Vibrant Matter: A Political Ecology of Things*. Durham, NC: Duke University Press.

Bentley, R. A., Maddison, E. J., Ranner, P. H., et al., 2014. Social tipping points and Earth systems dynamics. *Frontiers in Environmental Science* 2.

Berkes, F., 2008. *Sacred Ecology*, 2nd edn. New York: Routledge.

Beymer-Farris, B. A. and Bassett, T. J., 2012. The REDD menace: Resurgent protectionism in Tanzania's mangrove forests. *Global Environmental Change* 22(2): 332–41.

Bhattarai, B., Beilin, R., and Ford, R., 2015. Gender, agrobiodiversity, and climate change: A study of adaptation practices in the Nepal Himalayas. *World Development* 70: 122–32.

Biagini, B., Bierbaum, R., Stults, M., Dobardzic, S., and McNeeley, S. M., 2014. A typology of adaptation actions: A global look at climate adaptation actions financed through the Global Environment Facility. *Global Environmental Change* 25: 97–108.

Biermann, F., Bai, X., Bondre, N., et al., 2016. Down to earth: Contextualizing the Anthropocene. *Global Environmental Change* 39: 341–50.

Blake, E. S. and Zelinsky, D. A., 2018. National Hurricane Center Tropical Cyclone Report: Hurricane Harvey. National Oceanic & Atmospheric Administration and National Weather Service. Available at: https://www.nhc.noaa.gov/data/tcr/AL092017_Harvey.pdf [Accessed January 4, 2019].

Bloomberg News, 2017. China fossil fuel deadline shifts focus to electric car race. *Bloomberg.com*, September 10.

Bony, S., Stevens, B., Frierson, D. M. W., et al., 2015. Clouds, circulation and climate sensitivity. *Nature Geoscience* 8(4): 261–8.

Boston Consulting Group, 2018. The electric car tipping point future of powertrains for owned and shared mobility. *The Electric Car Tipping Point*. Available at: https://www.bcg.com/en-us/publications/2018/electric-car-tipping-point.aspx [Accessed February 20, 2018].

Boykoff, M. T., 2011. *Who Speaks for the Climate? Making Sense of Media Reporting on Climate Change*. Cambridge, UK: Cambridge University Press.

Bracken, L. J., Bulkeley, H. A., and Maynard, C. M., 2014. Micro-hydro power in the UK: The role of communities in an emerging energy resource. *Energy Policy* 68: 92–101.

Brady, M. B., 2018. *Mapping Coastal Exposure to Climate Risks in Alaska's North Slope: A Collaborative, Community-based Assessment*. PhD Dissertation, Rutgers University, NJ.

Bremer, S. and Meisch, S., 2017 Co-production in climate change research: Reviewing different perspectives. *Wiley Interdisciplinary Reviews: Climate Change* 8(6): e482.

Brondizio, E. S., O'Brien, K., Bai, X., et al., 2016. Re-conceptualizing the Anthropocene: A call for collaboration. *Global Environmental Change* 39: 318–27.

Brulle, R. J. and Dunlap, R., 2015. Sociology and global climate change: Introduction, in R. Dunlap and R. J. Brulle (eds), *Climate Change and Society: Sociological Perspectives*. New York: Oxford University Press, 1–31.

Bruvoll, A. and Larsen, B. M., 2004. Greenhouse gas emissions in Norway: Do carbon taxes work? *Energy Policy* 32(4): 493–505.

Buhaug, H., 2017. Climate, peace and security. *PRIO Climate and Conflict*. Available at: https://blogs.prio.org/ClimateAndConflict/2017/11/climate-peace-and-security/. [Accessed July 23, 2018]

Bulkeley, H., 2013. *Cities and Climate Change*. New York: Routledge.

Bulkeley, H., Carmin, J., Castán Broto, V., Edwards, G. A. S., and Fuller, S., 2013. Climate justice and global cities: Mapping the emerging discourses. *Global Environmental Change* 23(5): 914–25.

Butzer, K. W., 2012. Collapse, environment, and society. *Proceedings of the National Academy of Sciences* 109(10): 3632–9.

Butzer, K. W. and Endfield, G. H., 2012. Critical perspectives on historical collapse. *Proceedings of the National Academy of Sciences* 109(10): 3628–31.

Byers, M., Franks, K., and Gage, A., 2017. The internationalization of climate damages litigation. *Washington Journal of Environmental Law & Policy* 7: 264.

Cameron, E. S., 2012. Securing indigenous politics: A critique of the vulnerability and adaptation approach to the human dimensions of climate change in the Canadian Arctic. *Global Environmental Change* 22(1): 103–14.

Campbell, B. M., Thornton, P., Zougmoré, R., van Asten, P., and Lipper, L., 2014. Sustainable intensification: What is its role in climate smart agriculture? *Current Opinion in Environmental Sustainability* 8: 39–43.

Carabelli, G. and Lyon, D., 2016. Young people's orientations to the future: Navigating the present and imagining the future. *Journal of Youth Studies* 19(8): 1110–27.

CARE, 2016. Hope dries up? Women and girls coping with drought and climate change in Mozambique. Available at: https://careclimatechange.org/wp-content/uploads/2016/11/El_Nino_Mozambique_Report_final.pdf [Accessed July 23, 2018].

Carlsson-Kanyama, A. and González, A. D., 2009. Potential contributions of food consumption patterns to climate change. *The American Journal of Clinical Nutrition* 89(5): 1704S–9S.

Castán Broto, V., 2017. Urban governance and the politics of climate change. *World Development* 93: 1–15.

Castree, N., 2001. Socializing nature: Theory, practice, and politics, in N. Castree and B. Braun (eds), *Social Nature: Theory, Practice, and Politics*. Malden, MA: Blackwell, 1–21.

Castree, N., 2014. The Anthropocene and the environmental humanities: Extending the conversation. *Environmental Humanities* 5(1): 233–60.

CDP, 2018. CDP (formerly Climate Disclosure Project). Available at: https://www.cdp.net/en [Accessed May 25, 2018].

Centola, D., Becker, J., Brackbill, D., and Baronchelli, A., 2018. Experimental evidence for tipping points in social convention. *Science* 360(6393): 1116–19.

Chambwera, M., Heal, G., Dubeux, C., et al., 2014. Economics of adaptation, in

C. B. Field, V. R. Barros, D. J. Dokken, et al. (eds), *Climate Change 2014: Impacts, Adaptation, and Vulnerability. Part A: Global and Sectoral Aspects. Contribution of Working Group II to the Fifth Assessment Report of the Intergovernmental Panel on Climate Change*. Cambridge, UK: Cambridge University Press, 945–77.

Chapin III, F. S., Knapp, C. N., Brinkman, T. J., Bronen, R., and Cochran, P., 2016. Community-empowered adaptation for self-reliance. *Current Opinion in Environmental Sustainability* 19: 67–75.

Chapman, D. A., Lickel, B., and Markowitz, E. M., 2017. Reassessing emotion in climate change communication. *Nature Climate Change* 7(12): 850–2.

Chesters, G. and Welsh, I., 2005. Complexity and social movement(s): Process and emergence in planetary action systems. *Theory, Culture & Society* 22(5): 187–211.

Chu, E., Anguelovski, I., and Carmin, J., 2016. Inclusive approaches to urban climate adaptation planning and implementation in the global south. *Climate Policy* 16(3): 372–92.

City of Boston, 2018. Climate Ready Boston. *Boston.gov*. Available at: https://www.boston.gov/departments/environment/climate-ready-boston [Accessed May 15, 2018].

City of Copenhagen, 2011. Copenhagen Climate Adaptation Plan. Available at: http://en.klimatilpasning.dk/media/568851/copenhagen_adaption_plan.pdf [Accessed July 23, 2018].

Clapp, J., Newell, P., and Brent, Z. W., 2018. The global political economy of climate change, agriculture and food systems. *The Journal of Peasant Studies* 45(1): 80–8.

Clayton, S., Manning, C. M., Krygsman, K., and Speiser, M., 2017. *Mental Health and Our Changing Climate: Impacts, Implications and Guidance*. Washington, DC: American Psychological Association and ecoAmerica.

Cleveland, D.A., Carruth, A., and Mazaroli, D. N., 2015. Operationalizing local food: Goals, actions, and indicators for alternative food systems. *Agriculture and Human Values* 32(2): 281–97.

Coady, D., Parry, I., Sears, L., and Shang, B., 2017. How large are global fossil fuel subsidies? *World Development* 91: 11–27.

Cole, L. W. and Foster, S. R., 2001. *From the Ground Up: Environmental Racism and the Rise of the Environmental Justice Movement*. New York: New York University Press.

Coleman, J., 2007, Comments on Global Warming. Available at: http://ftpcontent.worldnow.com/kusi/Comments+on+Global+Warming02.pdf.

Commission on Human Security, 2003. *Human Security Now: Protecting and Empowering People*. New York: Commission on Human Security.

Conca, J. 2015. EROI: A tool to predict the best energy mix. *Forbes*. Available at: https://www.forbes.com/sites/jamesconca/2015/02/11/eroi-a-tool-to-predict-the-best-energy-mix/ [Accessed October 31, 2018].

Cook, J., Oreskes, N., Doran, P. T., et al., 2016. Consensus on consensus: A synthesis of consensus estimates on human-caused global warming. *Environmental Research Letters* 11(4): 048002.

Coole, D. and Frost, S. (eds), 2010. *New Materialisms: Ontology, Agency, and Politics*. Durham, NC: Duke University Press.

Costanza, R., de Groot, R., Sutton, P., et al., 2014. Changes in the global value of ecosystem services. *Global Environmental Change* 26: 152–8.

Cote, M. and Nightingale, A. J., 2012. Resilience thinking meets social theory:

Situating social change in socio-ecological systems (SES) research. *Progress in Human Geography* 36(4): 475–89.
Cunsolo, A. and Ellis, N. R., 2018. Ecological grief as a mental health response to climate change-related loss. *Nature Climate Change* 8(4): 275–81.
Cutter, S. L., Ash, K. D., and Emrich, C. T., 2014. The geographies of community disaster resilience. *Global Environmental Change* 29: 65–77.
Cutter, S. L., Boruff, B. J., and Shirley, W. L., 2003. Social vulnerability to environmental hazards. *Social Science Quarterly* 84(2): 242–61.
Dalby, S., 2016. Framing the Anthropocene: The good, the bad and the ugly. *The Anthropocene Review* 3(1): 33–51.
D'Ancona, M., 2017. *Post Truth: The New War on Truth and How to Fight Back*. London: Ebury Press.
Dangelico, R. M. and Vocalelli, D., 2017. "Green marketing": An analysis of definitions, strategy steps, and tools through a systematic review of the literature. *Journal of Cleaner Production* 165: 1263–79.
Daniel, S., 2011. Land grabbing and potential implications for world food security, in M. Behnassi, S. Shahid, and J. D'Silva (eds), *Sustainable Agricultural Development*. Dordrecht: Springer, 25–42.
Davies, K., Tabucanon, G. M., and Box, P., 2016. Children, climate change, and the intergenerational right to a viable future, in N. Ansell, N. Klocker, and Skelton, (eds), *Geographies of Global Issues: Change and Threat*. Singapore: Springer Singapore, 401–22.
Davis, M., 2006. *Planet of Slums*. London: Verso.
Davis, S. J. and Caldeira, K., 2010. Consumption-based accounting of CO_2 emissions. *Proceedings of the National Academy of Sciences of the United States of America*, 107(12): 5687–92.
Dean, J. F., Middelburg, J. J., Röckmann, T., et al., 2018. Methane feedbacks to the global climate system in a warmer world. *Reviews of Geophysics* 56(1): 207–50.
Dear Climate, 2018. Dear Climate. Available at: http://www.dearclimate.net/#/homepage [Accessed January 21, 2018].
Dearing, J. A., Wang, R., Zhang, K., et al., 2014. Safe and just operating spaces for regional social-ecological systems. *Global Environmental Change* 28: 227–38.
De Backer, K. and Miroudot, S., 2013. Mapping global value chains, *OECD Trade Policy Papers*, No. 159. Paris: OECD Publishing. Available at: https://doi.org/10.1787/5k3v1trgnbr4-en [Accessed October 31, 2018].
Demuzere, M., Orru, K., Heidrich, O., et al., 2014. Mitigating and adapting to climate change: Multi-functional and multi-scale assessment of green urban infrastructure. *Journal of Environmental Management* 146: 107–15.
Denton, F., 2002. Climate change vulnerability, impacts, and adaptation: Why does gender matter? *Gender & Development* 10(2): 10–20.
Denton, F., Wilbanks, T., Abeysinghe, A. C., et al., 2014. Climate-resilient pathways: Adaptation, mitigation, and sustainable development, in *Climate Change 2014: Impacts, Adaptation, and Vulnerability. Part A: Global and Sectoral Aspects. Contribution of Working Group II to the Fifth Assessment Report of the Intergovernmental Panel on Climate Change*. Cambridge, UK: Cambridge University Press, 1101–31.
Devall, B. and Sessions, G., 2001. *Deep Ecology: Living as if Nature Mattered*. Salt Lake City, UT: Gibbs Smith.

De Witt, A., 2015. Climate change and the clash of worldviews: An exploration of how to move forward in a polarized debate. *Zygon®* 50(4): 906–21.

Dhakal, S., 2010. GHG emissions from urbanization and opportunities for urban carbon mitigation. *Current Opinion in Environmental Sustainability* 2(4): 277–83.

Dicken, P., 2015. *Global Shift: Mapping the Changing Contours of the World Economy*, 7th edn. New York: Guilford Press.

Dobson, A., 2010. The fiction of climate change. *openDemocracy*. Available at: http://www.opendemocracy.net/andrew-dobson/fiction-of-climate-change [Accessed May 24, 2018].

Dodman, D., 2009. Blaming cities for climate change? An analysis of urban greenhouse gas emissions inventories. *Environment and Urbanization* 21(1): 185–201.

Doney, S. C., Fabry, V. J., Feely, R. A., and Kleypas, J. A., 2009. Ocean acidification: The other CO_2 problem. *Annual Review of Marine Science* 1(1): 169–92.

Dow, K., Berkhout, F., Preston, B. L., Klein, R. J. T., Midgley, G., and Shaw, M. R., 2013. Limits to adaptation. *Nature Climate Change* 3(4): 305–7.

Dryzek, J. S., 2013. *The Politics of the Earth: Environmental Discourses*, 3rd edn. Oxford: Oxford University Press.

Dwiartama, A. and Rosin, C., 2014. Exploring agency beyond humans: The compatibility of Actor-Network Theory (ANT) and resilience thinking. *Ecology and Society* 19(3): 28.

Eakin, H. and Luers, A. L., 2006. Assessing the vulnerability of social-environmental systems. *Annual Review of Environment and Resources* 31(1): 365–94.

Economist, 2011. A man-made world. Available at: https://www.economist.com/briefing/2011/05/26/a-man-made-world [Accessed October 25, 2018].

Edmondson, D., Chaudoir, S. R., Mills, M. A., Park, C. L., Holub, J., and Bartkowiak, J. M., 2011. From shattered assumptions to weakened worldviews: Trauma symptoms signal anxiety buffer disruption. *Journal of Loss & Trauma* 16(4): 358–85.

Edwards, F., Dixon, J., Friel, S., et al., 2011. Climate change adaptation at the intersection of food and health. *Asia Pacific Journal of Public Health* 23(2 suppl): 91S–104S.

Edwards, G. and Roberts, J. T., 2014. *A High-carbon Partnership? Chinese–Latin American Relations in a Carbon-constrained World*. Global Economy & Development, Brookings Institution, Working Paper 72.

EEA (European Environment Agency), 2017. *Introduction*. European Environment Agency, EEA Report No. 25. Copenhagen: EEA.

Ehrhardt-Martinez, K., et al., 2015. Climate change and consumption, in R. Dunlap and R. J. Brulle (eds), *Climate Change and Society: Sociological Perspectives*. New York: Oxford University Press, 93–126.

El Khoury, A., 2015. *Globalization Development and Social Justice: A Propositional Political Approach*. New York: Routledge.

Ellsworth, W. L., 2013. Injection-induced earthquakes. *Science* 341(6142): 1225942.

Ensor, J. and Berger, R., 2009. Community-based adaptation and culture in theory and practice, in W. N. Adger, I. Lorenzoni, and K. O'Brien (eds), *Adapting to Climate Change: Thresholds, Values, Governance*. Cambridge, UK: Cambridge University Press, 227–39.

Erdman, J., 2015. All-time record heat in Germany. *The Weather Channel*. Available at: https://weather.com/news/climate/news/europe-heat-wave-poland-germany-czech-august-2015 [Accessed March 4, 2018].

Eriksen, S., Aldunce, P., Bahinipati, C. S., et al., 2011. When not every response to climate change is a good one: Identifying principles for sustainable adaptation. *Climate and Development* 3(1): 7–20.

Eriksen, S., Inderberg, T. H., O'Brien, K., and Sygna, L., 2015. Introduction: Development as usual is not enough, in T. H. Inderberg, S. Eriksen, K. O'Brien, and L. Sygna (eds), *Climate Change Adaptation and Development: Transforming Paradigms and Practices*. New York: Routledge, 1–18.

Eriksen, S. H., Nightingale, A. J., and Eakin, H., 2015. Reframing adaptation: The political nature of climate change adaptation. *Global Environmental Change* 35: 523–33.

Erkens, G. and Sutanudjaja, E. H., 2015. Towards a global land subsidence map. *Proceedings of the International Association of Hydrological Sciences* 372: 83–7.

Escobar, A., 2015. Degrowth, postdevelopment, and transitions: A preliminary conversation. *Sustainability Science* 10(3): 451–62.

Falkner, R., 2016. The Paris Agreement and the new logic of international climate politics. *International Affairs* 92(5): 1107–25.

FAO (Food and Agriculture Organization of the United Nations), 2016. Climate-smart agriculture. Available at: http://www.fao.org/climate-smart-agriculture/overview/en/ [Accessed August 7, 2016].

FAO (Food and Agriculture Organization of the United Nations), 2018. SOFI 2017 – The State of Food Security and Nutrition in the World. Available at: http://www.fao.org/state-of-food-security-nutrition/en/ [Accessed March 23, 2018].

Farbotko, C. and Lazrus, H., 2012. The first climate refugees? Contesting global narratives of climate change in Tuvalu. *Global Environmental Change* 22(2): 382–90.

Farrell, A. D., Rhiney, K., Eitzinger, A., and Umaharan, P., 2018. Climate adaptation in a minor crop species: Is the cocoa breeding network prepared for climate change? *Agroecology and Sustainable Food Systems* 42(7): 812–33.

Fazey, I., 2010. Resilience and higher order thinking. *Ecology and Society* 15(3): 9.

Fazey, I., Moug, P., Allen, S., et al., 2018. Transformation in a changing climate: A research agenda. *Climate and Development* 10(3): 197–217.

Fazey, I., Wise, R. M., Lyon, C., Câmpeanu, C., Moug, P., and Davies, T. E., 2016. Past and future adaptation pathways. *Climate and Development* 8(1): 26–44.

Feola, G., 2015. Societal transformation in response to global environmental change: A review of emerging concepts. *Ambio* 44(5): 376–90.

Few, R., Brown, K., and Tompkins, E. L., 2007. Public participation and climate change adaptation: Avoiding the illusion of inclusion. *Climate Policy* 7(1): 46–59.

Fischer, E. M. and Knutti, R., 2015. Anthropogenic contribution to global occurrence of heavy-precipitation and high-temperature extremes. *Nature Climate Change* 5(6): 560–4.

Fischer, W., Hake, J.-F., Kuckshinrichs, W., Schröder, T., and Venghaus, S., 2016. German energy policy and the way to sustainability: Five controversial issues in the debate on the "Energiewende". *Energy* 115: 1580–91.

Flora, J. A. and Roser-Renouf, C., 2014. Climate change activism and youth, in UNICEF Office of Research: *The Challenges of Climate Change: Children on the Front Line*. Florence, Italy: Innocenti Insight, 86–91.

Ford, J. D., Stephenson, E., Willox, A. C., et al., 2016. Community-based adaptation research in the Canadian Arctic. *Wiley Interdisciplinary Reviews: Climate Change* 7(2): 175–91.

Fortune, 2018. Fortune Global 500 List 2017: See Who Made It. *Fortune*. Available at: http://fortune.com/global500/ [Accessed February 19, 2018].

Fouquet, R. and Pearson, P. J. G. 2006. Seven centuries of energy services: The price and use of light in the United Kingdom (1300–2000). *The Energy Journal* 27(1): 139–77.

Friedlingstein, P., Andrew, R. M., Rogelj, J., et al., 2014. Persistent growth of CO_2 emissions and implications for reaching climate targets. *Nature Geoscience* 7(10): 709–15.

Friedman, T. L., 2005. *The World is Flat: A Brief History of the Twenty-first Century*, 1st edn. New York: Farrar, Straus and Giroux.

Fritze, J. G., Blashki, G. A., Burke, S., and Wiseman, J., 2008. Hope, despair and transformation: Climate change and the promotion of mental health and wellbeing. *International Journal of Mental Health Systems* 2(1): 13.

Fujisawa, M., Kobayashi, K., Johnston, P., and New, M., 2015. What drives farmers to make top-down or bottom-up adaptation to climate change and fluctuations? A comparative study on 3 cases of apple farming in Japan and South Africa. *PLOS ONE* 10(3): e0120563.

Füssel, H.-M. and Klein, R. J. T., 2006. Climate change vulnerability assessments: An evolution of conceptual thinking. *Climatic Change* 75(3): 301–29.

Gabrys, J. and Yusoff, K., 2012. Arts, sciences and climate change: Practices and politics at the threshold. *Science as Culture* 21(1): 1–24.

Gain, A. K., Giupponi, C., and Wada, Y., 2016. Measuring global water security towards sustainable development goals. *Environmental Research Letters* 11(12): 124015.

Galafassi, D., Kagan, S., Milkoreit, M., et al., 2018. "Raising the temperature": The arts on a warming planet. *Current Opinion in Environmental Sustainability* 31: 71–9.

Gallucci, M., 2018. Rebuilding Puerto Rico's grid. *IEEE Spectrum* 55(5): 30–8.

Gardiner, S. M., 2004. Ethics and global climate change. *Ethics* 114(3): 555–600.

Gardiner, S. M., 2006. A perfect moral storm: Climate change, intergenerational ethics and the problem of moral corruption. *Environmental Values* 15(3): 397–413.

Gasparatos, A., Doll, C. N. H., Esteban, M., Ahmed, A., and Olang, T. A., 2017. Renewable energy and biodiversity: Implications for transitioning to a green economy. *Renewable and Sustainable Energy Reviews* 70: 161–84.

Gavin, M. C., McCarter, J., Mead, A., et al., 2015. Defining biocultural approaches to conservation. *Trends in Ecology & Evolution* 30(3): 140–5.

GEA, 2012. *Global Energy Assessment – Toward a Sustainable Future*. Vienna: International Institute for Applied Systems and Analysis and Cambridge, UK: Cambridge University Press.

Gemenne, F., 2011. Why the numbers don't add up: A review of estimates and predictions of people displaced by environmental changes. *Global Environmental Change* 21: S41–S49.

Gerber, J., 1997. Beyond dualism: The social construction of nature and the natural and social construction of human beings. *Progress in Human Geography* 21(1): 1–17.

Gertner, J., 2016. Should the United States save Tangier Island from oblivion? *The New York Times*, July 6.

Gewirtzman, J., Natson, S., Richards, J.-A., et al., 2018. Financing loss and damage: Reviewing options under the Warsaw International Mechanism. *Climate Policy* 18(8): 1076–86.

Ghosh, A., 2016. *The Great Derangement: Climate Change and the Unthinkable*. Chicago, IL: University of Chicago Press.

Gibson-Graham, J. K., 2011. A feminist project of belonging for the Anthropocene. *Gender, Place & Culture* 18(1): 1–21.

Gillis, J. and Fountain, H., 2016. Global warming cited as wildfires increase in fragile boreal forest. *The New York Times*, May 10.

Glaeser, E. L. and Kahn, M. E., 2010. The greenness of cities: Carbon dioxide emissions and urban development. *Journal of Urban Economics* 67(3): 404–18.

Global Carbon Project, 2017. Carbon budget and trends 2017. Available at: www.globalcarbonproject.org/carbonbudget [Accessed October 16, 2018].

Global Wind Energy Council, 2017. *Global Wind Report Annual Market Update 2016*. Washington, DC: Global Wind Energy Council.

Godfray, H. C. J., Beddington, J.R., Crute, I. R., et al., 2010. Food security: The challenge of feeding 9 billion people. *Science* 327(5967): 812–18.

González-Eguino, M., 2015. Energy poverty: An overview. *Renewable and Sustainable Energy Reviews* 47: 377–85.

Göpel, M., 2016. *The Great Mindshift: How a New Economic Paradigm and Sustainability Transformations Go Hand in Hand*. Cham, Switzerland: Springer International.

Gramberger, M., Zellmer, K., Kok, K., and Metzger, M. J., 2015. Stakeholder integrated research (STIR): A new approach tested in climate change adaptation research. *Climatic Change* 128(3–4): 201–14.

Grothmann, T. and Patt, A., 2005. Adaptive capacity and human cognition: The process of individual adaptation to climate change. *Global Environmental Change* 15(3): 199–213.

Grubb, M., Sha, F., Spencer, T., Hughes, N., Zhang, Z., and Agnolucci, P., 2015. A review of Chinese CO_2 emission projections to 2030: The role of economic structure and policy. *Climate Policy* 15 (suppl. 1): S7–S39.

Grün, G. 2015. Global climate change: Data-driven answers to the biggest questions. Available at: https://p.dw.com/p/1HC5Z [Accessed October 31, 2018].

Gunderson, L. H. and Holling, C. S. (eds), 2002. *Panarchy: Understanding Transformations in Human and Natural Systems*. Washington, DC: Island Press.

Hackmann, H., Moser, S. C., and St Clair, A. L., 2014. The social heart of global environmental change. *Nature Climate Change*. Available at: https://www.nature.com/articles/nclimate2320 [Accessed May 25, 2018].

Hall, C. A. S., Lambert, J. G., and Balogh, S. B., 2014. EROI of different fuels and the implications for society. *Energy Policy* 6: 141–52.

Hall, C. M., Baird, T., James, M., and Ram, Y., 2016. Climate change and cultural heritage: Conservation and heritage tourism in the Anthropocene. *Journal of Heritage Tourism* 11(1): 10–24.

Hallegatte, S., Green, C., Nicholls, R. J., and Corfee-Morlot, J., 2013. Future flood losses in major coastal cities. *Nature Climate Change* 3(9): 802–6.

Hampton, P., 2018. Trade unions and climate politics: Prisoners of neoliberalism or swords of climate justice? *Globalizations* 15(4): 470–86.

Hansen, J., Sato, M., and Ruedy, R., 2012. Perception of climate change. *Proceedings of the National Academy of Sciences* 109(37): E2415–23.

Haraway, D., 2015. Anthropocene, Capitalocene, Plantationocene, Chthulucene: Making kin. *Environmental Humanities* 6(1): 159–65.

Hargreaves, T., 2011. Practice-ing behaviour change: Applying social practice theory to pro-environmental behaviour change. *Journal of Consumer Culture* 11(1): 79–99.

Hastrup, K., 2016. A history of climate change: Inughuit responses to changing ice conditions in North-West Greenland. *Climatic Change*: 1–12.

Hawken, P., 2008. *Blessed Unrest: How the Largest Social Movement in History is Restoring Grace, Justice, and Beauty to the World*. New York: Penguin Books.

Hawken, P., 2017. *Drawdown: The Most Comprehensive Plan Ever Proposed to Reverse Global Warming*. New York: Penguin Books.

Head, L., 2016. *Hope and Grief in the Anthropocene: Re-conceptualising Human–Nature Relations*. New York: Routledge.

Hedlund-de Witt, A., 2013. Worldviews and their significance for the global sustainable development debate. *Environmental Ethics* 35(2): 133–62.

Hedlund-de Witt, A., 2014. The integrative worldview and its potential for sustainable societies: A qualitative exploration of the views and values of environmental leaders. *Worldviews: Global Religions, Culture, and Ecology* 18(3): 191–229.

Heffron, R. J. and McCauley, D., 2018. What is the "just transition"? *Geoforum* 88: 74–7.

Hegland, S. J., Nielsen, A., Lázaro, A., Bjerknes, A.-L., and Totland, Ø., 2009. How does climate warming affect plant–pollinator interactions? *Ecology Letters* 12(2): 184–95.

Henry, D. and Ramirez-Marquez, J. E., 2016. On the impacts of power outages during Hurricane Sandy: A resilience-based analysis. *Systems Engineering* 19(1): 59–75.

Hernandez, R. R., Easter, S. B., Murphy-Mariscal, M. L., et al., 2014. Environmental impacts of utility-scale solar energy. *Renewable and Sustainable Energy Reviews* 29: 766–79.

Heyd, T. and Brooks, N., 2009. Exploring cultural dimensions of adaptation to climate change, in W. N. Adger, I. Lorenzoni and K. O'Brien (eds), *Adapting to Climate Change: Thresholds, Values, Governance*. Cambridge, UK: Cambridge University Press, 269–82.

Higgins, P., Short, D., and South, N., 2013. Protecting the planet: A proposal for a law of ecocide. *Crime, Law and Social Change* 59(3): 251–66.

Hjerpe, M. and Linnér, B.-O., 2009. Utopian and dystopian thought in climate change science and policy. *Futures* 41(4): 234–45.

Hochschild, A., 2006. *Bury the Chains: Prophets and Rebels in the Fight to Free an Empire's Slaves*. Boston and New York: Mariner Books.

Hoffmann, A. A. and Sgrò, C. M., 2011. Climate change and evolutionary adaptation. *Nature* 470(7335): 479–85.

Holder, J., Ochagavia, E., Poulton, L., et al., 2015. The new cold war: Drilling for oil and gas in the Arctic. *The Guardian*. Available at: http://www.theguardian.com/environment/ng-interactive/2015/jun/16/drilling-oil-gas-arctic-alaska [Accessed February 17, 2018].

Holling, C., Gunderson, L. H., and Ludwig, D., 2002. In quest of a theory of adaptive change, in L. H. Gunderson and C. S. Holling (eds), *Panarchy: Understanding Transformations in Human and Natural Systems*. Washington, DC: Island Press, 3–24.

Hornsey, M. J., Harris, E. A., Bain, P. G., and Fielding, K. S., 2016. Meta-analyses of the determinants and outcomes of belief in climate change. *Nature Climate Change* 6(6): 622–6.

Hovelsrud, G. K. and Smit, B., 2010. *Community Adaptation and Vulnerability in Arctic Regions*. Dordrecht, Netherlands: Springer Science+Business Media.

Hughes, D. M., 2017. *Energy without Conscience: Oil, Climate Change, and Complicity*. Durham, NC: Duke University Press.

Hulme, M., 2009. *Why We Disagree about Climate Change: Understanding Controversy, Inaction and Opportunity*, 4th edn. Cambridge, UK: Cambridge University Press.

Hulme, M., 2011. Reducing the future to climate: A story of climate determinism and reductionism. *Osiris* 26(1): 245–66.

Hulme, M., 2014. Attributing weather extremes to "climate change": A review. *Progress in Physical Geography* 38(4): 499–511.

Hulme, M. (ed.), 2015. *Climates and Cultures*. Los Angeles, CA: SAGE Reference.

Hunter, P. R., 2003. Climate change and waterborne and vector-borne disease. *Journal of Applied Microbiology* 94(s1): 37–46.

Huynen, M. M. T. E., Martens, P., and Akin, S.-M., 2013. Climate change: An amplifier of existing health risks in developing countries. *Environment, Development and Sustainability* 15(6): 1425–42.

Hyde Park Progress, 2007. Urban Density & Climate Change. Available at: hydeparkprogress.blogspot.com [Accessed October 16, 2018].

Hyman, I. E. and Jalbert, M. C., 2017. Misinformation and worldviews in the post-truth information age: Commentary on Lewandowsky, Ecker, and Cook. *Journal of Applied Research in Memory and Cognition* 6(4): 377–81.

ILO (International Labour Organization), 2018. *World Employment and Social Outlook 2018: Greening with Jobs*. Geneva: ILO.

Inderberg, T. H., Eriksen, S., O'Brien, K., and Sygna, L., 2014. *Climate Change Adaptation and Development: Transforming Paradigms and Practices*. New York: Routledge.

Inglehart, R. and Welzel, C., 2010. Changing mass priorities: The link between modernization and democracy. *Perspectives on Politics* 8(2): 551–67.

International Energy Agency (IEA) and the World Bank, 2017. *Sustainable Energy for All 2017 – Progress toward Sustainable Energy*. Washington, DC: World Bank. Available at: https://trackingsdg7.esmap.org/data/files/download-documents/eegp17-01_gtf_full_report_for_web_0516.pdf [Accessed October 21, 2018].

International Energy Agency (IEA), 2018. The future of cooling: Opportunities for energy-efficient air conditioning. Available at: http://www.iea.org/publications/freepublications/publication/The_Future_of_Cooling.pdf [Accessed October 26, 2018].

IPCC, 1990. *Climate Change: The IPCC Scientific Assessment. Report Prepared for Intergovernmental Panel on Climate Change by Working Group I*. Cambridge, UK: Cambridge University Press.

IPCC, 2012. *Managing the Risks of Extreme Events and Disasters to Advance Climate Change Adaption: Special Report of the Intergovernmental Panel on Climate Change*. Cambridge, UK: Cambridge University Press.

IPCC, 2013a. Summary for policymakers, in T. F. Stocker, D. Qin, G.-K. Plattner, et al. (eds), *Climate Change 2013: The Physical Science Basis. Contribution of Working Group I to the Fifth Assessment Report of the Intergovernmental Panel on Climate Change*. Cambridge, UK: Cambridge University Press.

IPCC, 2013b. *Climate Change 2013: The Physical Science Basis. Contribution of Working Group I to the Fifth Assessment Report of the Intergovernmental Panel on Climate Change*. Cambridge, UK: Cambridge University Press.

IPCC, 2014a. *Climate Change 2014: Synthesis Report. Contribution of Working Groups I, II and III to the Fifth Assessment Report of the Intergovernmental Panel on Climate Change*. Geneva: IPCC.

IPCC, 2014b. Summary for policymakers, in C. B. Field, V. R. Barros, D. J. Dokken, et al. (eds), *Climate Change 2014: Impacts, Adaptation, and Vulnerability. Part A: Global and Sectoral Aspects. Contribution of Working Group II to the Fifth Assessment Report of the Intergovernmental Panel on Climate Change*. Cambridge, UK: Cambridge University Press, 1–32.

IPCC, 2014c. *Climate Change 2014: Mitigation of Climate Change: Working Group III contribution to the Fifth Assessment Report of the Intergovernmental Panel on Climate Change*. New York: Cambridge University Press.

IPCC, 2014d. Annex II: Glossary, in K. J. Mach, S. Planton and C. von Stechow (eds), *Climate Change 2014: Synthesis Report. Contribution of Working Groups I, II and III to the Fifth Assessment Report of the Intergovernmental Panel on Climate Change* . Geneva: IPCC, 117–30.

IPCC, 2018a. *Intergovernmental Panel on Climate Change: Organization*. Available at: https://www.ipcc.ch/organization/organization.shtml [Accessed July 19, 2018].

IPCC, 2018b *Global Warming of 1.5 °C: An IPCC Special Report on the Impacts of Global Warming of 1.5 °C above Pre-industrial Levels and Related Global Greenhouse Gas Emission Pathways, in the Context of Strengthening the Global Response to the Threat of Climate Change, Sustainable Development, and Efforts to Eradicate Poverty*. Available at: http://www.ipcc.ch/report/sr15/ [Accessed October 29, 2018].

IRENA (International Renewable Energy Agency), 2018. *Renewable Power Generation Costs in 2017*. Abu Dhabi: International Renewable Energy Agency.

Islam, M. and Kotani, K., 2016. Changing seasonality in Bangladesh. *Regional Environmental Change* 16(2): 585–90.

Islam, M., Sallu, S., Hubacek, K., and Paavola, J., 2014. Limits and barriers to adaptation to climate variability and change in Bangladeshi coastal fishing communities. *Marine Policy* 43: 208–16.

Jarosz, L., 2014. Comparing food security and food sovereignty discourses. *Dialogues in Human Geography* 4(2): 168–81.

Jeffers, J. M., 2013. Double exposures and decision making: Adaptation policy and planning in Ireland's coastal cities during a boom–bust cycle. *Environment and Planning A* 45(6): 1436–54.

Jenkins, K., McCauley, D., Heffron, R., Stephan, H., and Rehner, R., 2016. Energy justice: A conceptual review. *Energy Research & Social Science* 11: 174–82.

Jiang, P., Chen, Y., Geng, Y., et al., 2013. Analysis of the co-benefits of climate change mitigation and air pollution reduction in China. *Journal of Cleaner Production*, 58: 130–7.

Johnstone, S. and Mazo, J., 2011. Global warming and the Arab Spring. *Survival* 53(2): 11–17.

Jolly, W. M., Cochrane, M. A., Freeborn, P. H., et al., 2015. Climate-induced variations in global wildfire danger from 1979 to 2013. *Nature Communications* 6(1).

Jones, L. and Boyd, E., 2011. Exploring social barriers to adaptation: Insights from Western Nepal. *Global Environmental Change* 21(4): 1262–74.

Jones, M. D. and Peterson, H., 2017. Narrative persuasion and storytelling as climate communication strategies. Available at: http://climatescience.oxfordre.com/view/10.1093/acrefore/9780190228620.001.0001/acrefore-9780190228620-e-384 [Accessed July 23, 2018].

Jones, R. N., 2001. An environmental risk assessment/management framework for climate change impact assessments. *Natural Hazards* 23(2–3): 197–230.

Jones, R. N. and Preston, B.L., 2011. Adaptation and risk management. *Wiley Interdisciplinary Reviews: Climate Change* 2(2): 296–308.

Joule, E., 2011. Fashion-forward thinking: Sustainability as a business model at Levi Strauss. *Global Business and Organizational Excellence* 30 (2): 16–22.

Kahan, D., 2012. Why we are poles apart on climate change. *Nature News* 488(7411): 255.

Kahan, D., Peters, E., Wittlin, M., et al., 2012. The polarizing impact of science literacy and numeracy on perceived climate change risks. *Nature Climate Change* 2(10): 732–5.

Kahn, D. and Mulkern, A. C., 2017. Scientists see climate change in California's wildfires. *Scientific American*. Available at: https://www.scientificamerican.com/article/scientists-see-climate-change-in-californias-wildfires/ [Accessed February 26, 2018].

Kaika, D. and Zervas, E., 2013a. The environmental Kuznets curve (EKC) theory. Part A: Concept, causes and the CO_2 emissions case. *Energy Policy* 62 (Supplement C): 1392–1402.

Kaika, D. and Zervas, E., 2013b. The environmental Kuznets curve (EKC) theory. Part B: Critical issues. *Energy Policy* 62 (Supplement C): 1403–11.

Kaplan, E. A., 2015. *Climate Trauma: Foreseeing the Future in Dystopian Film and Fiction*. New Brunswick, NJ: Rutgers University Press.

Kegan, R. and Lahey, L. L., 2009. *Immunity to Change: How to Overcome It and Unlock the Potential in Yourself and Your Organization*. Boston, MA: Harvard Business Review Press.

Keil, K., 2013. The Arctic: A new region of conflict? The case of oil and gas. *Cooperation and Conflict* 49(2): 162–90.

Kenis, A. and Lievens, M., 2014. Searching for "the political" in environmental politics. *Environmental Politics* 23(4): 531–48.

Kennedy, J. J., Thorne, P. W., Peterson, T. C. et al., 2010: How do we know the world has warmed? in "State of the Climate in 2009." *Bulletin of the American Meteorological Society* 91(7): S79–S106.

Kiehl, J. T., 2016. *Facing Climate Change: An Integrated Path to the Future*. New York: Columbia University Press.

Killian, B., Rivera, L., Soto, M., and Navichoc, D., 2013. Carbon footprint across the coffee supply chain: The case of Costa Rican coffee. *Journal of Agricultural Science and Technology B*, 3(3B): 151.

Kim, H.-E., 2011. Changing climate, changing culture: Adding the climate change dimension to the protection of intangible cultural heritage. *International Journal of Cultural Property* 18(3): 259–90.

Kirby, P. and O'Mahony, T., 2018. *The Political Economy of the Low-Carbon Transition Pathways beyond Techno-Optimism*. Cham, Switzerland: Palgrave Macmillan.

Kittner, N. and Kammen, D. M., 2018. A battery of innovative choices – if we commit to investing. *Bulletin of the Atomic Scientists* 74(1): 7–10.

Klein, N., 2015. *This Changes Everything: Capitalism vs the Climate*. New York: Simon & Schuster.

Klinenberg, E., 2015. *Heat Wave: A Social Autopsy of Disaster in Chicago*, 2nd edn. Chicago, IL: University of Chicago Press.

Klinsky, S., Roberts, T., Huq, S., et al., 2017. Why equity is fundamental in climate change policy research. *Global Environmental Change* 44: 170–3.

Knight, K. W. and Schor, J. B., 2014. Economic growth and climate change: A cross-national analysis of territorial and consumption-based carbon emissions in high-income countries. *Sustainability* 6(6): 3722–31.

Knox-Hayes, J. K., 2016. *The Cultures of Markets: The Political Economy of Climate Governance*. New York: Oxford University Press.

Knutti, R., Rogelj, J., Sedláček, J., and Fischer, E. M., 2016. A scientific critique of the two-degree climate change target. *Nature Geoscience* 9(1): 13–18.

Koplitz, S. N., Mickley, L. J., Marlier, M. E., et al., 2016. Public health impacts of the severe haze in Equatorial Asia in September–October 2015: Demonstration of a new framework for informing fire management strategies to reduce downwind smoke exposure. *Environmental Research Letters* 11(9): 094023.

Kopp, R. E., DeConto, R. M., Bader, D. A., et al., 2017. Evolving understanding of Antarctic ice-sheet physics and ambiguity in probabilistic sea-level projections. *Earth's Future* 5(12): 2017EF000663.

Kreibich, H., Di Baldassarre, G., Vorogushyn, S., et al., 2017. Adaptation to flood risk: Results of international paired flood event studies. *Earth's Future* 5(10): 953–65.

Kruk, M. C., Parker, B., Marra, J. J., et al., 2017. Engaging with users of climate information and the coproduction of knowledge. *Weather, Climate, and Society* 9(4): 839–49.

Kuehne, G., 2014. How do farmers' climate change beliefs affect adaptation to climate change? *Society & Natural Resources* 27(5): 492–506.

Kuhn, T. S., 1962. *The Structure of Scientific Revolutions*, 2nd edn. Chicago, IL: University of Chicago Press.

Kuhns, R. J. and Shaw, G. H., 2018. *Navigating the Energy Maze: The Transition to a Sustainable Future*. Cham, Switzerland: Springer International.

Kunreuther, H. and Lyster, R., 2017. *The Role of Public and Private Insurance in Reducing Losses from Extreme Weather Events and Disasters*. Rochester, NY: Social Science Research Network, SSRN Scholarly Paper No. ID 2973656.

Lata, S. and Nunn, P., 2012. Misperceptions of climate-change risk as barriers to climate-change adaptation: A case study from the Rewa Delta, Fiji. *Climatic Change* 110(1–2): 169–86.

Leichenko, R., 2011. Climate change and urban resilience. *Current Opinion in Environmental Sustainability* 3(3): 164–8.

Leichenko, R., 2018. Vulnerable regions in a changing climate, in G. L. Clark, M. P. Feldman, M. S. Gertler, and D. Wójcik (eds), *The New Oxford Handbook of Economic Geography*. Available at: http://www.oxfordhandbooks.com/view/10.1093/oxfordhb/9780198755609.001.0001/oxfordhb-9780198755609-e-30 [Accessed July 23, 2018].

Leichenko, R. and O'Brien, K., 2008. *Environmental Change and Globalization: Double Exposures*. New York: Oxford University Press.

Leichenko, R. and Silva, J. A., 2014. Climate change and poverty: Vulnerability, impacts, and alleviation strategies. *Wiley Interdisciplinary Reviews: Climate Change* 5(4): 539–56.

Leichenko, R. and Solecki, W., 2005. Exporting the American Dream: The globalization of suburban consumption landscapes. *Regional Studies* 39(2): 241–53.

Leichenko, R., Major, D. C., Johnson, K., Patrick, L., and O'Grady, M., 2011. An economic analysis of climate change impacts and adaptations in New York State. *Annals of the New York Academy of Sciences* 1244 (Annex III): 1–145.

Leichenko, R., McDermott, M., Bezborodko, E., Brady, M., and Namendorf, E., 2014. Economic vulnerability to climate change in coastal New Jersey: A stakeholder-based assessment. *Journal of Extreme Events* 1(1): 1450003.

Leichenko, R., McDermott, M., and Bezborodko, E., 2015. Barriers, limits and limitations to resilience. *Journal of Extreme Events* 02(1): 1550002.

Leiserowitz, A., 2006. Climate change risk perception and policy preferences: The role of affect, imagery, and values. *Climatic Change* 77(1–2): 45–72.

Leiserowitz, A., Maibach, E., Roser-Renouf, C., Feinberg, G., and Rosenthal, S., 2015. *Climate Change in the American Mind: March, 2015*. New Haven, CT: Yale University and George Mason University, Yale Project on Climate Change Communication.

Lelieveld, J., Evans, J. S., Fnais, M., Giannadaki, D., and Pozzer, A., 2015. The contribution of outdoor air pollution sources to premature mortality on a global scale. *Nature* 525 (7569): 367–71.

Lenton, T. M., 2013. Environmental tipping points. *Annual Review of Environment and Resources* 38(1): 1–29.

Leonard, M., Westra, S., Phatak, A., et al., 2014. A compound event framework for understanding extreme impacts. *Wiley Interdisciplinary Reviews: Climate Change* 5(1): 113–28.

Levermann, A., Clark, P. U., Marzeion, B., et al., 2013. The multimillennial sea-level commitment of global warming. *Proceedings of the National Academy of Sciences of the United States of America* 110(34): 13745–50.

Lewis, S. L. and Maslin, M. A., 2015. Defining the Anthropocene. *Nature* 519(7542): 171–80.

Lieberoth, A., Holm Jensen, N., and Bredahl, T., 2018. Selective psychological effects of nudging, gamification and rational information in converting commuters from cars to buses: A controlled field experiment. *Transportation Research Part F: Traffic Psychology and Behaviour* 55: 246–61.

Liotta, P. H. and Owen, T., 2006. Why human security? *The Whitehead Journal of Diplomacy and International Relations*, 37–54.

Liu, W., Lund, H., Mathiesen, B. V., and Zhang, X., 2011. Potential of renewable energy systems in China. *Applied Energy* 88(2): 518–25.

Liverman, D., 2009. The geopolitics of climate change: Avoiding determinism, fostering sustainable development: An editorial comment. *Climatic Change* 96(1–2): 7–11.

Liverman, D. and Glasmeier, A., 2014. What are the economic consequences of climate change? *The Atlantic*, April 22. Available at: https://www.theatlantic.com/business/archive/2014/04/the-economic-case-for-acting-on-climate-change/360995/ [Accessed July 23, 2018].

Livezey, R. E., Vinnikov, K. Y., Timofeyeva, M. M., Tinker, R., and van den Dool, H. M., 2007. Estimation and extrapolation of climate normals and climatic trends. *Journal of Applied Meteorology and Climatology* 46(11): 1759–76.

Lorenzoni, I. and Hulme, M., 2009. Believing is seeing: Laypeople's views of future socio-economic and climate change in England and in Italy. *Public Understanding of Science* 18(4): 383–400.

Lövbrand, E., Beck, S., Chilvers, J., et al., 2015. Who speaks for the future of Earth? How critical social science can extend the conversation on the Anthropocene. *Global Environmental Change* 32: 211–18.

Lynch, A. H. and Veland, S., 2018. *Urgency in the Anthropocene*. Cambridge, MA: MIT Press.

Manyena, S., O'Brien, G., O'Keefe, P., and Rose, J., 2011. Disaster resilience: A bounce back or bounce forward ability? *Local Environment*, 16(5): 417–24.

Marcotullio, P. and McGranahan, G., 2007. *Scaling Urban Environmental Challenges: From Local to Global and Back*. London: Earthscan.

Marcotullio, P. J., Hughes, S., Sarzynski, A., et al., 2014. Urbanization and the carbon cycle: Contributions from social science. *Earth's Future* 2(10): 2014EF000257.

Marengo, J. A. and Espinoza, J. C., 2016. Extreme seasonal droughts and floods in Amazonia: Causes, trends and impacts. *International Journal of Climatology* 36(3): 1033–50.

Markham, A., Osipova, E., Lafrenz Samuels, K., and Caldas, A., 2016. *World Heritage and Tourism in a Changing Climate*. Paris: UNESCO Publishing.

Marquis, C., Toffel, M. W., and Zhou, Y., 2016. Scrutiny, norms, and selective disclosure: A global study of greenwashing. *Organization Science* 27(2): 483–504.

Marzeion, B. and Levermann, A., 2014. Loss of cultural world heritage and currently inhabited places to sea-level rise. *Environmental Research Letters* 9(3): 034001.

Masters, J., 2017. Historic heat wave sweeps Asia, the Middle East and Europe. *Weather Underground*, June 6.

Mastrandrea, M. D., Mach, K. J., Plattner, G.-K., et al., 2011. The IPCC AR5 guidance note on consistent treatment of uncertainties: A common approach across the working groups. *Climatic Change* 108(4): 675–91.

Maxmen, A., 2018. As Cape Town water crisis deepens, scientists prepare for "Day Zero." *Nature*. Available at: http://www.nature.com/articles/d41586-018-01134-x [Accessed March 26, 2018].

Maxwell, S. L., Fuller, R. A., Brooks, T. M., and Watson, J. E. M., 2016. Biodiversity: The ravages of guns, nets and bulldozers. *Nature News* 536(7615): 143.

McAlpine, C. A., Etter, A., Fearnside, P. M., Seabrook, L., and Laurance, W. F., 2009. Increasing world consumption of beef as a driver of regional and global change: A call for policy action based on evidence from Queensland (Australia), Colombia and Brazil. *Global Environmental Change* 19(1): 21–33.

McAneney, J., McAneney, D., Musulin, R., Walker, G., and Crompton, R., 2016. Government-sponsored natural disaster insurance pools: A view from down-under. *International Journal of Disaster Risk Reduction* 15: 1–9.

McDermott, M., Mahanty, S., and Schreckenberg, K., 2013. Examining equity: A multidimensional framework for assessing equity in payments for ecosystem services. *Environmental Science & Policy* 33: 416–27.

McGlade, C. and Ekins, P., 2015. The geographical distribution of fossil fuels unused when limiting global warming to 2°C. *Nature* 517(7533): 187–90.

McGranahan, G., Balk, D., and Anderson, B., 2007. The rising tide: Assessing the risks of climate change and human settlements in low elevation coastal zones. *Environment and Urbanization* 19(1): 17–37.

McGuffie, K. and Henderson-Sellers, A., 2005. *A Climate Modelling Primer*, 3rd edn. Chichester, UK: John Wiley & Sons.

McInerney, F. A. and Wing, S. L., 2011. The Paleocene–Eocene thermal maximum: A perturbation of carbon cycle, climate, and biosphere with implications for the future. *Annual Review of Earth and Planetary Sciences* 39(1): 489–516.

McKibben, B., 2012. Global warming's terrifying new math. *Rolling Stone*. Available at: https://www.rollingstone.com/politics/news/global-warmings-terrifying-new-math-20120719 [Accessed January 27, 2018].

McLaughlin, K. A., Berglund, P., Gruber, M. J., Kessler, R. C., Sampson, N. A., and Zaslavsky, A. M., 2011. Recovery from PTSD following Hurricane Katrina. *Depression and Anxiety* 28(6): 439–46.

Meadows, D. H., 1999. *Leverage Points: Places to Intervene in a System*. Norwich, VT: Sustainability Institute.

Meinshausen, M., Meinshausen, N., Hare, W., et al., 2009. Greenhouse-gas emission targets for limiting global warming to 2°C. *Nature* 458(7242): 1158–62.

Melvin, A. M., Larsen, P., Boehlert, B., et al., 2017. Climate change damages to Alaska public infrastructure and the economics of proactive adaptation. *Proceedings of the National Academy of Sciences* 114(2): E122–E131.

Merchant, C., 2005. *Radical Ecology: The Search for a Livable World*, 2nd edn. New York: Routledge.

Meyssignac, B., Fettweis, X., Chevrier, R., and Spada, G., 2017. Regional sea level changes for the twentieth and the twenty-first centuries induced by the regional variability in Greenland ice sheet surface mass loss. *Journal of Climate* 30(6): 2011–28.

Miao, C., Sun, Q., Kong, D., and Duan, Q., 2016. Record-breaking heat in northwest China in July 2015: Analysis of the severity and underlying causes. *Bulletin of the American Meteorological Society* 97(12): S97–S101.

Miles, M., 2010. Representing nature: Art and climate change. *Cultural Geographies* 17(1): 19–35.

Milkoreit, M., 2016. The promise of climate fiction, in P. Wapner and H. Elver (eds), *Reimagining Climate Change*. New York: Routledge, 171–91.

Milkoreit, M., 2017a. Pop-cultural mobilization: Deploying *Game of Thrones* to shift US climate change politics. *International Journal of Politics, Culture, and Society*: 1–22.

Milkoreit, M., 2017b. Imaginary politics: Climate change and making the future. *Elementa: Science of the Anthropocene* 5: 62.

Milly, P. C. D., Betancourt, J., Falkenmark, M., et al., 2008. Stationarity is dead: Whither water management? *Science* 319 (5863): 573–4.

Mitchell, D., Heaviside, C., Vardoulakis, S., et al., 2016. Attributing human mortality during extreme heat waves to anthropogenic climate change. *Environmental Research Letters* 11(7): 074006.

Mitchell, T., 2011. *Carbon Democracy: Political Power in the Age of Oil.* London: Verso.

Mohr, S. and Khan, O., 2015. 3D printing and its disruptive impacts on supply chains of the future. *Technology Innovation Management Review* 5(11): 6.

Mol, A. P. J. and Spaargaren, G., 2000. Ecological modernisation theory in debate: A review. *Environmental Politics* 9(1): 17–49.

Mooney, H. A., Duraiappah, A., and Larigauderie, A., 2013. Evolution of natural and social science interactions in global change research programs. *Proceedings of the National Academy of Sciences* 110 (Suppl. 1): 3665–72.

Moser, S. C., 2016. Reflections on climate change communication research and practice in the second decade of the 21st century: What more is there to say? *Wiley Interdisciplinary Reviews: Climate Change* 7(3): 345–69.

Moser, S. C. and Dilling, L., 2011. Communicating climate change: Closing the science–action gap. *The Oxford Handbook of Climate Change and Society*. Oxford: Oxford University Press, 161–74.

Moser, S. C. and Ekstrom, J. A., 2010. A framework to diagnose barriers to climate change adaptation. *Proceedings of the National Academy of Sciences* 107(51): 22026–31.

Mullan, M., Kingsmill, N., Agrawala, S., and Kramer, A. M., 2015. National adaptation planning: Lessons from OECD countries, in W. L. Filho (eds), *Handbook of Climate Change Adaptation*. Berlin: Springer, 1165–82.

Myers, S. S., Smith, M. R., Guth, S., et al. 2017. Climate change and global food systems: Potential impacts on food security and undernutrition. *Annual Review of Public Health* 38(1): 259–77.

Mysterud, A., Easterday, W. R., Stigum, V. M., Aas, A. B., Meisingset, E. L., and Viljugrein, H., 2016. Contrasting emergence of Lyme disease across ecosystems. *Nature Communications* 7: 11882.

Naess, A. and Rothenberg, D., 1989. *Ecology, Community and Lifestyle: Outline of an Ecosophy*. Cambridge, UK: Cambridge University Press.

Najafi, M. R., Zwiers, F. W., and Gillett, N. P., 2015. Attribution of Arctic temperature change to greenhouse-gas and aerosol influences. *Nature Climate Change* 5(3): 246–9.

Narasimhan, T. E., 2015. Chennai floods are world's 8th most expensive natural disaster in 2015. *Business Standard India*, December 11.

NAS (National Academies of Sciences, Engineering, and Medicine), 2016. *Attribution of Extreme Weather Events in the Context of Climate Change*. Washington, DC: The National Academies.

NASA (National Aeronautics and Space Administration), 2017. NOAA data show 2016 warmest year on record globally. *NASA*. Available at: http://www.nasa.gov/press-release/nasa-noaa-data-show-2016-warmest-year-on-record-globally [Accessed January 10, 2018].

NASA (National Aeronautics and Space Administration), 2018. Global surface temperature | NASA Global Climate Change. *Climate Change: Vital Signs of the Planet*. Available at: https://climate.nasa.gov/vital-signs/global-temperature [Accessed May 23, 2018].

Nauta, A. L., Heijmans, M. M. P. D., Blok, D., et al., 2015. Permafrost collapse after shrub removal shifts tundra ecosystem to a methane source. *Nature Climate Change* 5(1): 67–70.

Neslen, A., 2015. Pentagon to lose emissions exemption under Paris climate deal. *The Guardian*. Available at: http://www.theguardian.com/environment/2015/dec/14/pentagon-to-lose-emissions-exemption-under-paris-climate-deal [Accessed February 11, 2018].

Neumann, B., Vafeidis, A. T., Zimmermann, J., and Nicholls, R. J., 2015. Future coastal population growth and exposure to sea-level rise and coastal flooding: A global assessment. *PLOS ONE* 10(3): e0118571.

New, M., Liverman, D., Schroder, H., and Anderson, K., 2011. Introduction: Four degrees and beyond: The potential for a global temperature increase of four degrees and its implications. *Philosophical Transactions: Mathematical, Physical and Engineering Sciences* 369(1934): 6–19.

Nightingale, A. J., 2011. Bounding difference: Intersectionality and the material production of gender, caste, class and environment in Nepal. *Geoforum* 42(2): 153–62.

Nikoleris, A., Stripple, J., and Tenngart, P., 2017. Narrating climate futures: Shared socioeconomic pathways and literary fiction. *Climatic Change* 143(3–4): 307–19.

Nilsson, N. J., 2014. *Understanding Beliefs*. Cambridge, MA: MIT Press.

NOAA (National Oceanic & Atmospheric Administration), 2018. ESRL Global Monitoring Division – Global Greenhouse Gas Reference Network. Available at: https://www.esrl.noaa.gov/gmd/ccgg/trends/monthly.html [Accessed May 24, 2018].

Nordhaus, W. D., 2017. Revisiting the social cost of carbon. *Proceedings of the National Academy of Sciences* 114(7): 1518–23.

Norgaard, K. M., 2006. "We don't really want to know": Environmental justice and socially organized denial of global warming in Norway. *Organization & Environment* 19(3): 347–70.

Norgaard, K. M., 2011. *Living in Denial: Climate Change, Emotions, and Everyday Life*. Cambridge, MA: MIT Press.

Norwegian Petroleum, 2018. Employment in the petroleum industry. Available at: https://www.norskpetroleum.no/en/economy/employment/ [Accessed May 24, 2018].

Nyborg, K., Anderies, J. M., Dannenberg, A., et al., 2016. Social norms as solutions. *Science* 354(6308): 42–3.

O'Brien, K., 2010. Do values subjectively define the limits to climate change adaptation? in W. N. Adger, I. Lorenzoni, and K. O'Brien (eds), *Adapting to Climate Change: Thresholds, Values, Governance*. Cambridge, UK: Cambridge University Press, 164–80.

O'Brien, K., 2013. The courage to change: Adaptation from the inside-out, in S. C. Moser and M. Boykoff (eds), *Successful Adaptation to Climate Change: Linking Science and Practice in a Rapidly Changing World*. New York: Routledge, 306–20.

O'Brien, K., 2015. Political agency: The key to tackling climate change. *Science* 350(6265): 1170–1.

O'Brien, K., 2017. Climate change adaptation and social transformation, in D. Richardson, N. Castree, M. F. Goodchild, A. Kobayashi, W. Liu, and R.A. Marston (eds), *International Encyclopedia of Geography: People, the Earth, Environment and Technology*. Oxford, UK: John Wiley & Sons, 1–8.

O'Brien, K., 2018. Is the 1.5°C target possible? Exploring the three spheres of transformation. *Current Opinion in Environmental Sustainability* 31: 153–60.

O'Brien, K. and Barnett, J., 2013. Global environmental change and human security. *Annual Review of Environment and Resources* 38(1): 373–91.

O'Brien, K. and Hochachka, G., 2010. Integral adaptation to climate change. *Journal of Integral Theory and Practice* 5(1): 89–102.

O'Brien, K. and Leichenko, R., 2006. Climate change, equity and human security. *Die Erde* 137(3): 223–40.

O'Brien, K. and Selboe, E., 2015. Social transformation: The real adaptive challenge, in K. O'Brien and E. Selboe (eds), *The Adaptive Challenge of Climate Change*. Cambridge, UK: Cambridge University Press, 311–24.

O'Brien, K. and Sygna, L. 2013. Responding to climate change: the three spheres of transformation, in Proceedings of Transformation in a Changing Climate Conference, University of Oslo, Norway, 16–23.

O'Brien, K., Eriksen, S., Nygaard, L. P., and Schjolden, A., 2007. Why different interpretations of vulnerability matter in climate change discourses. *Climate Policy* 7(1): 73–88.

O'Brien, K., Eriksen, S., Sygna, L., and Naess, L. O., 2006. Questioning complacency: Climate change impacts, vulnerability, and adaptation in Norway. *AMBIO: A Journal of the Human Environment* 35(2): 50–6.

O'Brien, K., Selboe, E., and Hayward, B. 2018. Exploring youth activism on climate change: Dutiful, disruptive and dangerous dissent. *Ecology and Society* 23(3): 42.

O'Brien, K., Sygna, L., and Haugen, J. E., 2004. Vulnerable or resilient? A multi-scale assessment of climate impacts and vulnerability in Norway. *Climatic Change* 64(1–2): 193–225.

Ojala, M., 2012. Hope and climate change: The importance of hope for environmental engagement among young people. *Environmental Education Research* 18(5): 625–42.

Olsson, P., Moore, M.-L., Westley, F., and McCarthy, D., 2017. The concept of the Anthropocene as a game-changer: A new context for social innovation and transformations to sustainability. *Ecology and Society* 22(2): 31.

O'Neill, B. C., Oppenheimer, M., Warren, R., et al., 2017. IPCC reasons for concern regarding climate change risks. *Nature Climate Change* 7(1): 28–37.

O'Neill, S., 2018. Engaging people with climate change imagery, in Matthew C. Nisbet, Shirley S. Ho, Ezra Markowitz, Saffron O'Neill, Mike S. Schäfer, and Jagadish Thaker (eds), *The Oxford Encyclopedia of Climate Change Communication*. Oxford University Press, online edn. Available at: http://climatescience.oxfordre.com/view/10.1093/acrefore/9780190228620.001.0001/acrefore-9780190228620-e-371 [Accessed October 31, 2018].

Oppenheimer, M., Campos, R., Warren, R., et al., 2014. Emergent risks and key vulnerabilities, in C. B. Field et al. (eds), *Climate Change 2014: Impacts, Adaptation, and Vulnerability. Part A: Global and Sectoral Aspects. Contribution of Working Group II to the Fifth Assessment Report of the Intergovernmental Panel on Climate Change.* Cambridge, UK: Cambridge University Press, 1039–109.

Oreskes, N. and Conway, E. M., 2010. *Merchants of Doubt: How a Handful of Scientists Obscured the Truth on Issues from Tobacco Smoke to Global Warming.* New York: Bloomsbury Press.

O'Riordan, T. and Jordan, A., 1999. Institutions, climate change and cultural theory: Towards a common analytical framework. *Global Environmental Change* 9(2): 81–93.

Orlove, B., 2005. Human adaptation to climate change: A review of three historical cases and some general perspectives. *Environmental Science & Policy* 8(6): 589–600.

Oslo Municipality, 2018. *Klimabudsjett 2018 (Climate Budget 2018).* Oslo, Norway.

Otto, F. E. L., van der Wiel, K., van Oldenborgh, et al., 2018. Climate change increases the probability of heavy rains in Northern England/Southern Scotland like those of Storm Desmond: A real-time event attribution revisited. *Environmental Research Letters* 13(2): 024006.

Oxfam, 2015. Extreme carbon inequality. Oxfam Media Briefing, December 2, 2015. Available at: https://www.oxfam.org/sites/www.oxfam.org/files/file_attachments/mb-extreme-carbon-inequality-021215-en.pdf [Accessed October 31, 2018].

Palsson, G., Szerszynski, B., Sörlin, S., et al., 2013. Reconceptualizing the "Anthropos" in the Anthropocene: Integrating the social sciences and humanities in global environmental change research. *Environmental Science & Policy* 28: 3–13.

Panayotou, T., 2003. Economic growth and the environment. *Economic Survey of Europe* 2: 45–72.

Pandey, D., Agrawal, M., and Pandey, J. S., 2011. Carbon footprint: Current methods of estimation. *Environmental Monitoring and Assessment* 178(1–4): 135–60.

Parham, P. E., Waldock, J., Christophides, G. K., et al., 2015. Climate, environmental and socio-economic change: Weighing up the balance in vector-borne disease transmission. *Philosophical Transactions of the Royal Society B: Biological Sciences* 370(1665): 20130551. Available at: http://dx.doi.org/10.1098/rstb.2013.0551 [Accessed October 31, 2018].

Patterson, J., Schulz, K., Vervoort, J., et al., 2017. Exploring the governance and politics of transformations towards sustainability. *Environmental Innovation and Societal Transitions* 24: 1–16.

Pearce, F., 2016. After Paris, a move to rein in emissions by ships and planes. *Yale E360.* Available at: http://e360.yale.edu/features/reduce_co2_emissions_shipping_aviation_regulation_paris [Accessed February 11, 2018].

Peek, L., 2008. Children and disasters: Understanding vulnerability, developing capacities, and promoting resilience: An introduction. *Children, Youth and Environments* 18(1): 1–29.

Pelling, M., 2011. *Adaptation to Climate Change: From Resilience to Transformation.* New York: Routledge.

Perdan, S. and Azapagic, A., 2011. Carbon trading: Current schemes and future developments. *Energy Policy* 39(10): 6040–54.

Perwaiz, A., 2015. Thailand floods and impact on private sector, in T. Izumi and

R. Shaw (eds), *Disaster Management and Private Sectors*. Tokyo: Springer Japan, 231–45.

Peters, G. P., Minx, J. C., Weber, C. L., and Edenhofer, O., 2011. Growth in emission transfers via international trade from 1990 to 2008. *Proceedings of the National Academy of Sciences* 108(21): 8903–8.

Petersen, A., Hals, H., Rot, B., et al., 2014. Climate change and the Jamestown S'Klallam tribe: A customized approach to climate vulnerability and adaptation planning. *Michigan Journal of Sustainability* 2 (20170719). Available at: https://doi.org/10.3998/mjs.12333712.0002.003 [Accessed October 31, 2018].

Phan, M. D., Montz, B. E., Curtis, S., and Rickenbach, T. M., 2018. Weather on the go: An assessment of smartphone mobile weather applications use among college students. *Bulletin of the American Meteorological Society* 99(11): 2245–57.

Pirard, P., Vandentorren, S., Pascal, M., et al., 2005. Summary of the mortality impact assessment of the 2003 heat wave in France. *Eurosurveillance* 10(7): 7–8.

Pope Francis 2015. *Encyclical Letter Laudato Si' of the Holy Father Francis: On Care for Our Common Home*. London: Catholic Truth Society.

Prati, G. and Zani, B., 2013. The effect of the Fukushima nuclear accident on risk perception, antinuclear behavioral intentions, attitude, trust, environmental beliefs, and values. *Environment and Behavior* 45(6): 782–98.

Primack, R. B., Higuchi, H., and Miller-Rushing, A. J., 2009. The impact of climate change on cherry trees and other species in Japan. *Biological Conservation* 142(9): 1943–9.

Princen, T., 2003. Principles for sustainability: From cooperation and efficiency to sufficiency. *Global Environmental Politics* 3(1): 33–50.

Prudham, S., 2009. Pimping climate change: Richard Branson, global warming, and the performance of green capitalism. *Environment and Planning A: Economy and Space* 41(7): 1594–1613.

Raffa, K. F., Aukema, B. H., Bentz, B. J., Carroll, A. L., Hicke, J. A., and Kolb, T. E., 2015. Responses of tree-killing bark beetles to a changing climate, in C. Björkman and P. Niemelä (eds), *Climate Change and Insect Pests*. Wallingford, UK: CABI, 173–201.

Ravetz, J. R., 2006. Post-normal science and the complexity of transitions towards sustainability. *Ecological Complexity* 3(4): 275–84.

Reese, A., 2018. As countries crank up the AC, emissions of potent greenhouse gases are likely to skyrocket. *Science* 359(6380): 1084.

REN21, 2017. *Renewables 2017 Global Status Report*. Paris: REN21 Secretariat.

Renn, O. and Marshall, J. P., 2016. Coal, nuclear and renewable energy policies in Germany: From the 1950s to the "Energiewende". *Energy Policy* 99: 224–32.

Ribot, J., 2014. Cause and response: Vulnerability and climate in the Anthropocene. *The Journal of Peasant Studies* 41(5): 667–705.

Rickards, L., Ison, R., Fünfgeld, H., and Wiseman, J., 2014. Opening and closing the future: Climate change, adaptation, and scenario planning. *Environment and Planning C: Politics and Space* 32(4): 587–602.

Riedy, C., 2009. The influence of futures work on public policy and sustainability. *Foresight* 11(5): 40–56.

Risser, M. D. and Wehner, M. F., 2017. Attributable human-induced changes in the likelihood and magnitude of the observed extreme precipitation during Hurricane Harvey. *Geophysical Research Letters* 44(24): 2017GL075888.

Ritchie, H. and Roser, M. 2018. CO_2 and other greenhouse gas emissions. Published online at OurWorldInData.org. Retrieved from https://ourworldindata.org/co2-and-other-greenhouse-gas-emissions.

Rockström, J., Steffen, W., Noone, K., et al., 2009. A safe operating space for humanity. *Nature*. Available at: https://www.nature.com/articles/461472a [Accessed May 21, 2018].

Rojas-Downing, M. M., Nejadhashemi, A .P., Harrigan, T., and Woznicki, S. A., 2017. Climate change and livestock: Impacts, adaptation, and mitigation. *Climate Risk Management* 16: 145–63.

Rosa, E. A. and Dietz, T., 2012. Human drivers of national greenhouse-gas emissions. *Nature Climate Change* 2(8): 581–6.

Rosenberg, R., 2018. Vital signs: Trends in reported vectorborne disease cases – United States and territories, 2004–2016. *MMWR. Morbidity and Mortality Weekly Report* 67(17): 496–501.

Rosenzweig, C., Solecki, W., Romero-Lankao, P., Mehrotra, S., Dhakal, S., and Ibrahim, S. A. (eds), 2018. *Climate Change and Cities: Second Assessment Report of the Urban Climate Change Research Network*. Cambridge, UK: Cambridge University Press.

Rothman, D. S., van Bers, C., Bakkes, J., and Pahl-Wostl, C., 2009. How to make global assessments more effective: Lessons from the assessment community. *Current Opinion in Environmental Sustainability* 1(2): 214–18.

Rubens, G. Z. de, Noel, L., and Sovacool, B. K., 2018. Dismissive and deceptive car dealerships create barriers to electric vehicle adoption at the point of sale. *Nature Energy* 3(6): 501–7.

Rulli, M. C., Saviori, A., and D'Odorico, P., 2013. Global land and water grabbing. *Proceedings of the National Academy of Sciences* 110(3): 892–7.

Russo, S., Marchese, A. F., Sillmann, J., and Immé, G., 2016. When will unusual heat waves become normal in a warming Africa? *Environmental Research Letters* 11(5): 054016.

Ryan, K., 2016. Incorporating emotional geography into climate change research: A case study in Londonderry, Vermont, USA. *Emotion, Space and Society* 19: 5–12.

Saadi, S., Todorovic, M., Tanasijevic, L., Pereira, L. S., Pizzigalli, C., and Lionello, P., 2015. Climate change and Mediterranean agriculture: Impacts on winter wheat and tomato crop evapotranspiration, irrigation requirements and yield. *Agricultural Water Management* 147: 103–15.

Sack, K., 2018. Left to Louisiana's tides, a village fights for time. *The New York Times*, February 24.

Salama, S. and Aboukoura, K., 2018. Role of emotions in climate change communication, in W. Leal Filho et al. (eds), *Handbook of Climate Change Communication: Vol. 1*. Cham, Switzerland: Springer International, 137–50.

Samenow, J., 2018. Arctic temperatures soar 45 degrees above normal, flooded by extremely mild air on all sides. *Washington Post*, February 22.

Sandstrom, S. and Juhola, S., 2017. Continue to blame it on the rain? Conceptualization of drought and failure of food systems in the Greater Horn of Africa. *Environmental Hazards* 16(1): 71–91.

Sarzynski, A., 2015. Public participation, civic capacity, and climate change adaptation in cities. *Urban Climate* 14: 52–67.

Satterthwaite, D., 2009. The implications of population growth and urbanization for climate change. *Environment and Urbanization* 21(2): 545–67.

Saunois, M., Jackson, R. B., Bousquet, P., Poulter, B., and Canadell, J. G., 2016. The growing role of methane in anthropogenic climate change. *Environmental Research Letters* 11(12): 120207.

Schleussner, C.-F., Rogelj, J., Schaeffer, M., et al., 2016. Science and policy characteristics of the Paris Agreement temperature goal. *Nature Climate Change* 6(9): 827–35.

Schlitz, M. M., Vieten, C., and Miller, E. M., 2010. Worldview transformation and the development of social consciousness. *Journal of Consciousness Studies* 17(7–8): 18–36.

Schlosberg, D., 2004. Reconceiving environmental justice: Global movements and political theories. *Environmental Politics* 13(3): 517–40.

Schlosberg, D., Collins, L. B., and Niemeyer, S., 2017. Adaptation policy and community discourse: Risk, vulnerability, and just transformation. *Environmental Politics* 26(3): 413–37.

Schmidt, J. J., Brown, P. G., and Orr, C. J., 2016. Ethics in the Anthropocene: A research agenda. *The Anthropocene Review* 3(3): 188–200.

Schubert, C., 2017. Green nudges: Do they work? Are they ethical? *Ecological Economics* 132: 329–42.

Schulte, D. M., Dridge, K. M., and Hudgins, M. H., 2015. Climate change and the evolution and fate of the Tangier Islands of Chesapeake Bay, USA. *Scientific Reports* 5: 17890.

Schwartz, S. H., 2007. Basic human values: Theory, methods, and application. *Risorsa Uomo* 2: 2–83.

Schwarz, M. and Thompson, M., 1990. *Divided We Stand: Redefining Politics, Technology, and Social Choice*. London: Harvester Wheatsheaf.

Sengupta, S., 2018. Hotter, drier, hungrier: How global warming punishes the world's poorest. *The New York Times*, March 12.

Serrao-Neumann, S., Crick, F., Harman, B., Schuch, G., and Choy, D. L., 2015. Maximising synergies between disaster risk reduction and climate change adaptation: Potential enablers for improved planning outcomes. *Environmental Science & Policy* 50: 46–61.

Seto, K. C., Davis, S. J., Mitchell, R. B., Stokes, E. C., Unruh, G., and Ürge-Vorsatz, D., 2016. Carbon lock-in: Types, causes, and policy implications. *Annual Review of Environment and Resources* 41(1): 425–52.

Seto, K. C., Guneralp, B., and Hutyra, L. R., 2012. Global forecasts of urban expansion to 2030 and direct impacts on biodiversity and carbon pools. *Proceedings of the National Academy of Sciences* 109(40): 16083–8.

Settele, J., Bishop, J., and Potts, S. G., 2016. Climate change impacts on pollination. *Nature Plants*. Available at: https://www.nature.com/articles/nplants201692 [Accessed February 24, 2018].

Seymour, F. and Busch, J., 2016. *Why Forests? Why Now? The Science, Economics and Politics of Tropical Forests and Climate Change*. Washington, DC: Center for Global Development.

Sharma, M., 2017. *Radical Transformational Leadership: Strategic Action for Change Agents*. Berkeley, CA: North Atlantic Books.

Shaw, A., Burch, S., Kristensen, F., Robinson, J., and Dale, A., 2014. Accelerating the sustainability transition: Exploring synergies between adaptation and mitigation in British Columbian communities. *Global Environmental Change* 25: 41–51.

Sheehan, P., Cheng, E., English, A., and Sun, F., 2014. China's response to the air pollution shock. *Nature Climate Change* 4: 306–9.

Sherwood, S. C. and Huber, M., 2010. An adaptability limit to climate change due to heat stress. *Proceedings of the National Academy of Sciences* 107(21): 9552–5.

Shove, E., 2010. Beyond the ABC: Climate change policy and theories of social change. *Environment and Planning A: Economy and Space* 42(6): 1273–85.

Sider, A. and Matthews, C. M., 2017. Henry Hub emerges as global natural gas benchmark. *Wall Street Journal*, August 17.

Sieff, K., 2018. As Cape Town's water runs out, the rich drill wells. The poor worry about eating. *Washington Post*, 23 February.

Siegrist, M., Sütterlin, B., and Keller, C., 2014. Why have some people changed their attitudes toward nuclear power after the accident in Fukushima? *Energy Policy* 69: 356–63.

Sklair, L., 2002. *Globalization: Capitalism and Its Alternatives*, 3rd edn. Oxford, UK: Oxford University Press.

Smit, B. and Wandel, J., 2006. Adaptation, adaptive capacity and vulnerability. *Global Environmental Change* 16(3): 282–92.

Smit, B., Burton, I., Klein, R. J. T., and Wandel, J., 2000. An anatomy of adaptation to climate change and variability. *Climatic Change* 45(1): 223–51.

Smith, R. and Sweet, C., 2013. Companies unplug from the electric grid, delivering a jolt to utilities. *Wall Street Journal*, September 18.

Soito, J. L. da S. and Freitas, M. A. V., 2011. Amazon and the expansion of hydropower in Brazil: Vulnerability, impacts and possibilities for adaptation to global climate change. *Renewable and Sustainable Energy Reviews* 15(6): 3165–77.

Solecki, W., Leichenko, R., and Eisenhauer, D., 2017. Extreme climate events, household decision-making and transitions in the immediate aftermath of Hurricane Sandy. *Miscellanea Geographica* 21(4): 139–50.

Solecki, W., Pelling, M., and Garschagen, M., 2017. Transitions between risk management regimes in cities. *Ecology and Society* 22(2): 38.

Sorrell, S., 2009. Jevons' paradox revisited: The evidence for backfire from improved energy efficiency. *Energy Policy* 37(4): 1456–69.

Springmann, M., Godfray, H. C. J., Rayner, M., and Scarborough, P., 2016. Analysis and valuation of the health and climate change cobenefits of dietary change. *Proceedings of the National Academy of Sciences* 113(15): 4146–51.

Standal, K. and Winther, T., 2016. Empowerment through energy? Impact of electricity on care work practices and gender relations. *Forum for Development Studies* 43(1): 27–45.

Steffen, W., Broadgate, W., Deutsch, L., Gaffney, O., and Ludwig, C., 2015a. The trajectory of the Anthropocene: The Great Acceleration. *The Anthropocene Review* 2(1): 81–98.

Steffen, W., Crutzen, P. J., and McNeill, J. R., 2007. The Anthropocene: Are humans now overwhelming the great forces of nature? *Ambio* 36(8): 614–21.

Steffen, W., Richardson, K., Rockstrom, J., et al., 2015b. Planetary boundaries: Guiding human development on a changing planet. *Science* 347(6223): 1259855-2–1259855-10.

Steffen, W., Rockström, J., Richardson, K., Lenton, T. M., Foke, C., Liverman, D. et al., 2018. Trajectories of the Earth system in the Anthropocene. *Proceedings of the National Academy of Science* 115(33): 8252–9.

Steffen, W., Sanderson, A., Tyson, P. D., et al., 2005. *Global Change and the Earth System: A Planet under Pressure*. Berlin and New York: Springer.

Steiger, R., Scott, D., Abegg, B., Pons, M., and Aall, C., 2017. A critical review of climate change risk for ski tourism. *Current Issues in Tourism*: 1–37.

Steinbruner, J. D., Stern, P.C., Husbands, J. L., and National Research Council (US), (eds), 2013. *Climate and Social Stress: Implications for Security Analysis*. Washington, DC: National Academies Press.

Stern, N., 2013. The structure of economic modeling of the potential impacts of climate change: Grafting gross underestimation of risk onto already narrow science models. *Journal of Economic Literature* 51(3): 838–59.

Stern, P. C., Perkins, J. H., Sparks, R. E., and Knox, R. A., 2016. The challenge of climate-change neoskepticism. *Science* 353(6300): 653–4.

Stevenson, E. G. J., Greene, L. E., Maes, K. C., et al. 2012. Water insecurity in 3 dimensions: An anthropological perspective on water and women's psychosocial distress in Ethiopia. *Social Science & Medicine* 75(2): 392–400.

Stewart, E. J., Howell, S. E. L., Draper, D., Yackel, J., and Tivy, A., 2007. Sea ice in Canada's Arctic: Implications for cruise tourism. *Arctic* 60(4): 370–80.

Stoknes, P. E., 2015. *What We Think about When We Try Not to Think about Global Warming: Toward a New Psychology of Climate Action*. White River Junction, VT: Chelsea Green Publishing.

Strauss, S., 2012. Are cultures endangered by climate change? Yes, but . . . *Wiley Interdisciplinary Reviews: Climate Change* 3(4): 371–7.

Svoboda, M., 2016. Cli-fi on the screen(s): Patterns in the representations of climate change in fictional films. *Wiley Interdisciplinary Reviews: Climate Change* 7(1): 43–64.

Sweet, W., Kopp, R., Weaver, C., et al., 2017. *Global and Regional Sea-Level Rise Scenarios for the United States*. Silver Spring, MD: National Oceanic and Atmospheric Administration, NOAA Technical Report NOS CO-OPS 083.

Sweet, W., Park, J., Marra, J., Zervas, C., and Gill, S., 2014. *Sea Level Rise and Nuisance Flood Frequency Changes around the United States*. Silver Spring, MD: National Oceanic and Atmospheric Administration, NOAA Technical Report NOS CO-OPS 073.

Swilling, M. and Annecke, E., 2012. *Just Transitions: Explorations of Sustainability in an Unfair World*. Claremont: UCT-Press.

Swim, J. K., Stern, P. C., Doherty, T. J., et al., 2011. Psychology's contributions to understanding and addressing global climate change. *American Psychologist* 66(4): 241–50.

Swyngedouw, E., 2013. The non-political politics of climate change. *An International Journal for Critical Geographies* 12: 1–8.

Tang, Q., Zhang, X., and Francis, J. A., 2014. Extreme summer weather in northern mid-latitudes linked to a vanishing cryosphere. *Nature Climate Change* 4(1): 45–50.

Teller-Elsberg, J., Sovacool, B., Smith, T., and Laine, E., 2016. Fuel poverty, excess winter deaths, and energy costs in Vermont: Burdensome for whom? *Energy Policy* 90: 81–91.

Thomas, A. and Leichenko, R., 2011. Adaptation through insurance: Lessons from the NFIP. *International Journal of Climate Change Strategies and Management* 3(3): 250–63.

Torres, R. D. C. S., Andrade, C., and Gomes, S. M. D. S., 2017. Construction of greenhouse gas inventory for a Brazilian distribution electricity company. *Latin American Journal of Management for Sustainable Development* 3(4): 261–88.

Trenberth, K. E., 2011. Changes in precipitation with climate change. *Climate Research* 47(1–2): 123–38.

Trentmann, F., 2016. *Empire of Things: How We Became a World of Consumers, from the Fifteenth Century to the Twenty-First.* New York: HarperCollins.

Tripathi, A. and Mishra, A. K., 2017. Knowledge and passive adaptation to climate change: An example from Indian farmers. *Climate Risk Management* 16: 195–207.

Tripati, A. K., Roberts, C. D., and Eagle, R. A., 2009. Coupling of CO2 and ice sheet stability over major climate transitions of the last 20 million years. *Science* 326(5958): 1394–7.

Tschakert, P. and Dietrich, K. A., 2010. Anticipatory learning for climate change adaptation and resilience. *Ecology and Society* 15(2): 11.

Tsoutsos, T., Frantzeskaki, N., and Gekas, V., 2005. Environmental impacts from the solar energy technologies. *Energy Policy* 33(3): 289–96.

UNFCCC (United Nations Framework Convention on Climate Change), 2015. Paris Agreement. Presented at the 21st Conference of the Parties, Paris: United Nations.

UNFCCC (United Nations Framework Convention on Climate Change), 2018. Warsaw International Mechanism for Loss and Damage. Available at: http://unfccc.int/adaptation/workstreams/loss_and_damage/items/8134.php [Accessed April 1, 2018].

UNGC (United Nations Global Compact). 2018. The Ten Principles of the UN Global Compact. Available at: https://www.unglobalcompact.org/what-is-gc/mission/principles [Accessed May 4, 2018].

United Kingdom Environment Agency, 2018. Estimating the economic costs of the 2015 to 2016 winter floods. Bristol, UK: Environment Agency.

United Nations, 2015. Sustainable development goals – United Nations. *United Nations Sustainable Development.* Available at: https://www.un.org/sustainabledevelopment/sustainable-development-goals/ [Accessed June 6, 2018].

United Nations, 2018. *World Urbanization Prospects: The 2018 Revision.* UN, Population Division, Department of Economic and Social Affairs.

United Nations Development Programme (ed.), 1994. *Human Development Report 1994.* New York: Oxford University Press.

United Nations Environment Programme, 2018. Why does green economy matter? *UN Environment.* Available at: http://www.unenvironment.org/explore-topics/green-economy/why-does-green-economy-matter [Accessed July 2, 2018].

United Nations General Assembly, 2012. *The Future We Want.* A/RES/66/288. 66th Session, Agenda item 19. Resolution adopted by the General Assembly on July 27. Available at: http://www.un.org/en/development/desa/population/migration/

generalassembly/docs/globalcompact/A_RES_66_288.pdf [Accessed October 31, 2018].

United States Department of Energy, 2017. *US Energy and Employment Report.* Available at: https://www.energy.gov/sites/prod/files/2017/01/f34/2017%20US%20Energy%20and%20Jobs%20Report_0.pdf [Accessed October 31, 2018].

United States Energy Information Administration, 2017. *International Energy Outlook 2017.* Washington, DC: United States Energy Information Administration. Available at: https://www.eia.gov/outlooks/ieo/pdf/0484(2017).pdf [Accessed October 31, 2018].

Unruh, G. C., 2000. Understanding carbon lock-in. *Energy Policy* 28(12): 817–30.

Urban, M. C., 2015. Accelerating extinction risk from climate change. *Science* 348(6234): 571–3.

USEPA (United States Environmental Protection Agency), 2012. *Climate Change Indicators in the United States*, 2nd edn, Washington, DC, USA: US EPA, p. 3. EPA 430-R-12-004. Available at: https://commons.wikimedia.org/wiki/File:Earth%27s_greenhouse_effect_(US_EPA,_2012).png

USEPA (United States Environmental Protection Agency), 2016. Understanding Global Warming Potentials. US EPA. Available at: https://www.epa.gov/ghgemissions/understanding-global-warming-potentials [Accessed January 9, 2018].

USGCRP, 2016. *The Impacts of Climate Change on Human Health in the United States: A Scientific Assessment.* Washington, DC: US Global Change Research Program,

USGCRP, 2017. *Climate Science Special Report: Fourth National Climate Assessment*, Vol. I. Washington, DC: US Global Change Research Program.

van Benthem, A. A., 2015. Energy leapfrogging. *Journal of the Association of Environmental and Resource Economists* 2(1): 93–132.

van der Land, V., Romankiewicz, C., and van der Geest, K., 2018. Environmental change and migration: A review of West African case studies, in R. McLeman and F. Gemenne (eds), *Routledge Handbook of Environmental Displacement and Migration.* New York: Routledge, 163–77.

van Eijck, J., Romijn, H., Balkema, A., and Faaij, A., 2014. Global experience with jatropha cultivation for bioenergy: An assessment of socio-economic and environmental aspects. *Renewable and Sustainable Energy Reviews* 32: 869–89.

van Renssen, S., 2017. The visceral climate experience. *Nature Climate Change* 7(3): 168–71.

van Renssen, S., 2018. The inconvenient truth of failed climate policies. *Nature Climate Change* 8(5): 355–8.

Veland, S., Scoville-Simonds, M., Gram-Hanssen, I., et al., 2018. Narrative matters for sustainability: The transformative role of storytelling in realizing 1.5°C futures. *Current Opinion in Environmental Sustainability* 31: 41–7.

Vermeulen, S. J., Campbell, B. M., and Ingram, J. S. I., 2012. Climate change and food systems. *Annual Review of Environment and Resources* 37(1): 195–222.

Vidic, R. D., Brantley, S. L., Vandenbossche, J. M., Yoxtheimer, D., and Abad, J. D., 2013. Impact of shale gas development on regional water quality. *Science* 340(6134): 1235009.

Wackernagel, M., Kitzes, J., Moran, D., Goldfinger, S., and Thomas, M., 2006. The ecological footprint of cities and regions: Comparing resource availability with resource demand. *Environment and Urbanization* 18(1): 103–12.

Warner, K. and van der Geest, K., 2013. Loss and damage from climate change: Local-level evidence from nine vulnerable countries. *International Journal of Global Warming* 5(4): 367–86.

Warner, R., 2010. Ecological modernisation theory: Towards a critical ecopolitics of change? *Environmental Politics* 19(4): 538–56.

Waters, C. N., Zalasiewicz, J., Summerhayes, C., et al., 2016. The Anthropocene is functionally and stratigraphically distinct from the Holocene. *Science* 351(6269): aad2622.

Watts, M., 2018. The resource curse, in N. Castree, M. Hulme, and J. D. Proctor (eds), *Companion to Environmental Studies*. New York: Routledge, 95–9.

WCED, 1987. *Our Common Future*. (The Brundtland Report). World Commission on Environment and Development. New York: Oxford University Press.

Weber, A., 2013. *Enlivenment: Towards a Fundamental Shift in the Concepts of Nature, Culture and Politics*. Berlin: Heinrich-Böll-Stiftung.

Weber, E. U., 2010. What shapes perceptions of climate change? *Wiley Interdisciplinary Reviews: Climate Change* 1(3): 332–42.

Weißbach, D., Ruprecht, G., Huke, A., Czerski, K., Gottlieb, S., Hussein, A., Energy intensities, EROIs (energy returned on invested), and energy payback times of electricity generating power plants, *Energy*, Volume 52, 2013, pp. 210–21.

Wendt, A., 2015. *Quantum Mind and Social Science: Unifying Physical and Social Ontology*. Cambridge, UK and New York: Cambridge University Press.

Westley, F. R., Tjornbo, O., Schultz, L., et al., 2013. A theory of transformative agency in linked social-ecological systems. *Ecology and Society* 18(3): 27.

Wheeler, T. and Braun, J. von, 2013. Climate change impacts on global food security. *Science* 341(6145): 508–13.

WHO (World Health Organization), 2018. Frequently asked questions. *WHO*. Available at: http://www.who.int/suggestions/faq/en/ [Accessed March 31, 2018].

WHO (World Health Organization), 2005. Priority risks maps. *WHO*. 2005. Available at: http://www.who.int/heli/risks/risksmaps/en/index1.html [Accessed October 25, 2018].

Wilbanks, T. J. and Kates, R. W., 2010. Beyond adapting to climate change: Embedding adaptation in responses to multiple threats and stresses. *Annals of the Association of American Geographers* 100(4): 719–28.

Wilber, K., 1999. *The Marriage of Sense and Soul: Integrating Science and Religion*. New York: Harmony.

Wilber, K., 2000. *A Brief History of Everything*. Boston, MA: Shambhala Publications.

Wilhite, H., 2016. *The Political Economy of Low Carbon Transformation: Breaking the Habits of Capitalism*. New York: Routledge.

Wise, R. M., Fazey, I., Stafford Smith, M., et al., 2014. Reconceptualising adaptation to climate change as part of pathways of change and response. *Global Environmental Change* 28: 325–36.

Wisner, B., Blaikie, P., Cannon, T., and Davis, I., 2003. *At Risk: Natural Hazards, People's Vulnerability and Disasters*, 2nd edn. London: Routledge.

WMO (World Meteorological Organization), 2017. *WMO Greenhouse Gas Bulletin: The State of Greenhouse Gases in the Atmosphere Based on Global Observations through 2016 (13)*. Geneva: WMO.

Wolfe, D. W., Comstock, J., Lasko, A., et al., 2011. Agriculture. *Annals of the New York Academy of Sciences* 1244: 217–54.

Wong, K., 2001. Mother nature's medicine cabinet. *Scientific American*. Available at: https://www.scientificamerican.com/article/mother-natures-medicine-c/ [Accessed October 16, 2018].

Wood, E. M. and Kellermann, J. L. (eds), 2015. *Phenological Synchrony and Bird Migration: Changing Climate and Seasonal Resources in North America*. Boca Raton, FL: CRC Press.

Workman, C. L. and Ureksoy, H., 2017. Water insecurity in a syndemic context: Understanding the psycho-emotional stress of water insecurity in Lesotho, Africa. *Social Science & Medicine* 179: 52–60.

World Bank, 2010. *World Development Report 2010: Development and Climate Change*. Washington, DC: World Bank.

World Bank, 2015. CO_2 emissions (metric tons per capita) data. Available at: https://data.worldbank.org/indicator/EN.ATM.CO2E.PC [Accessed February 4, 2018].

World Bank, 2018. World Bank open data. Available at: data.worldbank.org.

Wright, C. and Nyberg, D., 2016. An inconvenient truth: How organizations translate climate change into business as usual. *Academy of Management Journal* 60(5): 1633–61.

Wright, E. O., 2010. *Envisioning Real Utopias*. London and New York: Verso.

WTO (World Trade Organization), 2015. *International Trade Statistics 2015*. Geneva: WTO. Available at: https://www.wto.org/english/res_e/statis_e/its2015_e/its15_toc_e.htm

Wu, X., Lu, Y., Zhou, S., Chen, L., and Xu, B., 2016. Impact of climate change on human infectious diseases: Empirical evidence and human adaptation. *Environment International* 86: 14–23.

Wynes, S. and Nicholas, K. A., 2017. The climate mitigation gap: Education and government recommendations miss the most effective individual actions. *Environmental Research Letters* 12(7): 074024.

Xie, J., Sreenivasan, S., Korniss, G., Zhang, W., Lim, C., and Szymanski, B. K., 2011. Social consensus through the influence of committed minorities. *Physical Review E* 84(1): 011130.

Yohe, G. and Leichenko, R., 2010. Adopting a risk-based approach. *Annals of the New York Academy of Sciences* 1196: 29–40.

Zalasiewicz, J., Waters, C. N., Summerhayes, C. P., et al., 2017. The working group on the Anthropocene: Summary of evidence and interim recommendations. *Anthropocene* 19: 55–60.

Zhang, X., Li, H.-Y., Deng, Z. D., et al., 2018. Impacts of climate change, policy and water–energy–food nexus on hydropower development. *Renewable Energy* 116: 827–34.

Zhao, C., Liu, B., Piao, S., et al., 2017. Temperature increase reduces global yields of major crops in four independent estimates. *Proceedings of the National Academy of Sciences* 114(35): 9326–31.

Ziervogel, G., Cowen, A., and Ziniades, J., 2016. Moving from adaptive to transformative capacity: Building foundations for inclusive, thriving, and regenerative urban settlements. *Sustainability* 8(9): 955.

Ziervogel, G., Shale, M., and Du, M., 2010. Climate change adaptation in a developing country context: The case of urban water supply in Cape Town. *Climate and Development* 2(2): 94–110.

Ziervogel, G., Waddell, J., Smit, W., and Taylor, A., 2016. Flooding in Cape Town's informal settlements: Barriers to collaborative urban risk governance. *South African Geographical Journal* 98(1): 1–20.

Zografos, C., Anguelovski, I., and Grigorova, M., 2016. When exposure to climate change is not enough: Exploring heatwave adaptive capacity of a multi-ethnic, low-income urban community in Australia. *Urban Climate* 17: 248–65.

Zoomers, A., 2010. Globalisation and the foreignisation of space: Seven processes driving the current global land grab. *The Journal of Peasant Studies* 37(2): 429–47.

Zscheischler, J. and Seneviratne, S. I., 2017. Dependence of drivers affects risks associated with compound events. *Science Advances* 3(6): e1700263.

Index